Mechanisms and Machine Science

Volume 26

Series editor

Marco Ceccarelli, Cassino, Italy

More information about this series at http://www.springer.com/series/8779

Andrés Kecskeméthy · Francisco Geu Flores
Editors

Interdisciplinary Applications of Kinematics

Proceedings of the International Conference, Lima, Peru, September 9–11, 2013

Springer

Editors
Andrés Kecskeméthy
Francisco Geu Flores
Department of Mechanical Engineering
University of Duisburg-Essen
Duisburg
Germany

ISSN 2211-0984 ISSN 2211-0992 (electronic)
ISBN 978-3-319-34420-1 ISBN 978-3-319-10723-3 (eBook)
DOI 10.1007/978-3-319-10723-3

Springer Cham Heidelberg New York Dordrecht London

Printed on acid-free paper

Springer is part of Springer Science+Business Media (www.springer.com)

Preface

The present proceedings collects 19 papers that were selected after peer review for the Second Conference on Interdisciplinary Applications in Kinematics, held in Lima, Peru, during September 9–11, 2013.

The objective of the conference was to bring together researchers from different fields where kinematics plays a key role. This includes not only theoretical fields where kinematics is traditionally established, but also fields in which kinematics might contribute to new perspectives for practical applications. Examples are the areas of biomechanics, industrial machinery, molecular kinematics, railway vehicles, among many others.

The participation of 33 researchers and 10 students from 10 countries showed the strong interest these topics find in the scientific community. The site of the conference in Peru not only proved to be very successful, but also helped to foster the international scientific cooperation in this region, which has outstanding potentials.

We thank the authors for submitting their valuable contributions for this conference as well as the reviewers for performing the reviews in due time. We also thank the publisher Springer for the timely implementation of this book and the valuable advices during the production process. We are grateful to the Universidad de Piura Campus Lima, the Pontificia Universidad Católica del Perú as well as the University of Duisburg-Essen for sponsoring this conference and contributing to its success. Our special acknowledgments go to the Förderverein Ingenieurwissenschaften Universität Duisburg-Essen (Association of Friends of Engineering Science of the University of Duisburg-Essen) for their valuable contribution to the funding of the present proceedings. Last but not the least, we thank the International Federation for the Promotion of Mechanism and Machine Science IFToMM for the ideal support by offering its patronage for this conference.

Andrés Kecskeméthy
Francisco Geu Flores

Organization

Chair

Andrés Kecskeméthy (Germany)

Co-chair

Francisco Geu Flores (Germany)

International Scientific Committee

J. Ambrosio (Portugal)
J. Angeles (Canada)
M. Ceccarelli (Italy)
B. Corves (Germany)
J. Cuadrado (Spain)
S. Dubowsky (USA)
M. Dutra (Brasil)
P. Hagedorn (Germany)
B. Heimann (Germany)
J. Herder (The Netherlands)
M. Hiller (Germany)
T. Huang (China-Beijing)
M. Husty (Austria)
K. Kazerounian (USA)
J. Lenarcic (Slovenia)
C. Lopez-Cajun (Mexico)

W. Kowalczyk (Germany)
J.-P. Merlet (France)
M. Morlock (Germany)
A. Müller (Germany)
V. Parenti-Castelli (Italy)
A. Pott (Germany)
D. Pisla (Romania)
B. Roth (USA)
B. Siciliano (Italy)
P. Wenger (France)
C. Woernle (Germany)

Local Organizing Committee

R. Callupe
E. Carrera
D. Elías
R. Hernani

Patronage

IFToMM Perú
International Federation for the Promotion of Mechanism and Machine Science
University of Duisburg-Essen
Pontificia Universidad Católica del Perú
Universidad de Piura, Campus Lima

Contents

Operation Modes in Lower-Mobility Parallel Manipulators

Josef Schadlbauer, Latifah Nurahmi, Manfred Husty, Philippe Wenger and Stéphane Caro

Abstract This paper discusses the notion of operation mode in parallel manipulators with less than six dof. This notion has been reported recently in several papers but the physical meaning of an operation mode is not always clear. Indeed, even if in some cases an operation mode can be associated with an understandable motion type e.g., three pure translations or a spherical motion, in some other cases, such an association is not straightforward. Therefore, the axodes are used in this paper to characterize any operation mode of lower-mobility parallel manipulators. A 3-RPS manipulator is used as an illustrative example. This manipulator is special because one can parameterize its operation modes.

Keywords Lower-mobility parallel manipulators · Operation modes · Axodes

1 Introduction

Lower-mobility parallel manipulators (LMPMs) have less than six degrees of freedom (dof). They are useful in tasks that do not require full rigid motions of the end-effector. Typical LMPMs are those generating three pure translations (for, e.g. 3-axis machining), three pure rotations (spherical manipulators), Schönflies motions,

J. Schadlbauer · M. Husty
Institute for Basic Sciences in Engineering, Unit for Geometry and CAD,
University of Innsbruck, Innsbruck, Austria
e-mail: josef.schadlbauer@uibk.ac.at

M. Husty
e-mail: manfred.husty@uibk.ac.at

L. Nurahmi · P. Wenger (✉) · S. Caro
Institut de Recherche en Communications et Cybernétique de Nantes, Nantes, France
e-mail: philippe.wenger@irccyn.ec-nantes.fr

L. Nurahmi
e-mail: latifah.nurahmi@irccyn.ec-nantes.fr

S. Caro
e-mail: stephane.caro@irccyn.ec-nantes.fr

A. Kecskeméthy and F. Geu Flores (eds.), *Interdisciplinary Applications of Kinematics*,
Mechanisms and Machine Science 26, DOI 10.1007/978-3-319-10723-3_1

i.e. three translations and one rotation about an axis of fixed direction (e.g. for pick and place tasks in the food industry). A well known LMPM is the 3-RPS LMPM, originally introduced by Hunt (1983). This LMPM can produce one translation and two rotations at least locally. The 3-RPS LMPM has then been studied for several applications, such as in telescope applications (Carretero et al. 1997) and in tool heads of machine tools (Hernandez et al. 2008). The 3-RPS manipulator is used as an illustrative example in this paper. The moving platform of the 3-RPS LMPM is connected to a fixed base through three legs with one actuator per leg. Unlike 6-dof parallel manipulators in which all six legs have six dof, the legs of a fully-parallel LMPM may have more dof than the moving platform. The combination of the leg constraints generates a set of constraints on the moving platform, which produce its desired reduced mobility. However, in some configurations, referred to as constraint singularities (CS), the wrench system associated with the passive constraints of the legs degenerates, resulting in extra infinitesimal degrees of freedom of the moving platform (Zlatanov et al. 2002). Another interesting feature may exist in some LMPMs, which is intimately connected to the existence of CS. This feature is the existence of several operation modes, which is the main topic of this paper.

Next section recalls the link between local and global kinematic analyses. Section 3 explains the notion of operation mode. Several ways to interpret this notion are given. Last section is devoted to the motion analysis of a 3-RPS LMPM. For more simplicity, we do not consider LMPMs with multiple working modes and there is no leg singularity.

2 Local Versus Global Kinematic Analysis

A key point in the kinematic analysis of LMPMs is the level at which the kinematic behaviour is analyzed. At the local level, one is interested in determining the instantaneous motion of the Moving Platform (MP) with respect to its base. In a given configuration, screw theory (Ball 1900; Hunt 1978) can be used to determine the kinetostatic behaviour of a LMPM by determining the constraint and actuation wrenches applied on the MP. The wrench graph and the twist graph are useful tools to visualize these screw systems in the projective space (Amine et al. 2012). The motion of the MP is thus described locally by the twist system of the MP in the configuration at hand. Displacement group theory can also help us analyze the available local motions (Huynh and Hervé 2005). Starting from a local analysis, determining the motion of the MP in a more global way is not straightforward. In general, the order of the twist system of the MP does not change until a singularity is encountered. If it is so, the configuration is then called a full cycle configuration (Conconi and Carricato 2009) that defines a reference state in a full singularity-free region of the workspace of the LMPM at hand. In another singularity-free region, the LMPM may move in a distinct full cycle configuration. A complete global analysis of the kinematic behaviour of the LMPM requires the description and characterization of all its admissible full cycle configurations. This starts with the determination of all its direct kinematic solutions, as recalled below.

3 Operation Modes

To the best of our knowledge, the notion of operation modes (OM) was first pointed out in Zlatanov et al. (2002), where a LMPM called DYMO robot was shown to exhibit five distinct 3-dof OM, each one being associated with a particular type of platform motion: translational, rotational, planar (2 types) and mixed motions. In the aforementioned paper, the existence of multiple OM in a LMPM was intimately related to a kind of "multi-functionality". This multi-functionality appears as an interesting feature for the designer, as manipulators exhibiting this property could be used for a wider class of tasks. Note that a close but yet different notion is the notion of kinematotropic mechanisms introduced by Wohlhart in 1996 (Wohlhart 1996). It pertains to closed-loop mechanisms that exhibit different numbers of dof. This property will not be studied here. Surprisingly, little research work has been devoted to the study of OM since 2002. Fanghella et al. (2006) proposed some LMPM possessing different OM associated with distinct displacement groups with the same dof (e.g. planar motions and a 3-d of subset of Schnflies motions) or distinct variants of this displacement group (e.g. one Schönflies motion with a rotation about a vertical axis and another one with a rotation about an horizontal axis) (Fanghella et al. 2006). More recently, Walter et al. (2009) deeply investigated the OM of two 3-UPU LMPM: the Tsai 3-UPU (Walter and Husty 2011) and the SNU 3-UPU (Walter et al. 2009). Resorting to efficient algebraic geometry tools, they described for the first time all different types of motions that these manipulators exhibit in addition to their usual translational mode. Similarly, the 3-RPS LMPM was also deeply analyzed in 2011 (Schadlbauer and Husty 2011). The 3-RPS was shown to have two distinct OM but contrary to the abovementioned two UPU robots, the types of motion associated with these two OM cannot be easily distinguished.

3.1 Operation Modes Versus Assembly Modes

The word assembly mode (AM) was originally defined with reference to the way of assembling a given parallel manipulator for given actuated joint values. An AM is currently used to define a solution to the direct kinematic problem. The workspace is divided into several singularity-free regions, named "aspects" in Wenger and Chablat (1998). As a matter of fact, an OM is associated with several distinct AM. In the workspace, the OM form regions that collect several aspects. The OM are separated by constraint singularities.

3.2 Mathematical Description and Geometric Interpretation

Using Study coordinates $(x_0, x_1, x_2, x_3, y_0, y_1, y_2, y_3)$ (see e.g. Husty et al. 2007) to describe the spatial Euclidean displacements, a complete analysis of the motion capabilities of a special 3-R<u>P</u>S PM was performed in Schadlbauer and Husty (2011). The 3-R<u>P</u>S PM is a 3-dof PM proposed by Hunt (1983). This LMPM is often referred to as "1T2R manipulator" but this designation does not mean that the 3-RPS LMPM has two rotational degrees of freedom about two fixed axes.

The 3-R<u>P</u>S PM, shown in Fig. 1, is composed of three limbs $L_i = \mathsf{R}^i\underline{\mathsf{P}}^i\mathsf{S}^i$, $i = 1, 2, 3$ such that: (i) The axis of R^i is directed along $\mathbf{f}_{ic}$ such that $\mathbf{f}_{1c}$, $\mathbf{f}_{2c}$ and $\mathbf{f}_{3c}$ are three independent unit vectors parallel to the plane of the fixed base, namely, orthogonal to $\mathbf{n}$, (ii) The $\underline{\mathsf{P}}^i$-joint is directed along $\mathbf{f}_{ia}$, (iii) The S^i-joint is centered at point B_i, (iv) points A_i, B_i are vertices of equilateral triangles and $\mathbf{f}_{ic}$ have directions tangent to the circumcircle of A_i. In Schadlbauer and Husty (2011) it was found out that the workspace splits into two different components that are characterized by either $x_0 = 0$ or $x_1 = 0$. In Eq. (1) the set of constraint equations for the component $x_0 = 0$ is displayed.[1] The set for the other component is equally simple but is omitted here because of lack of space.

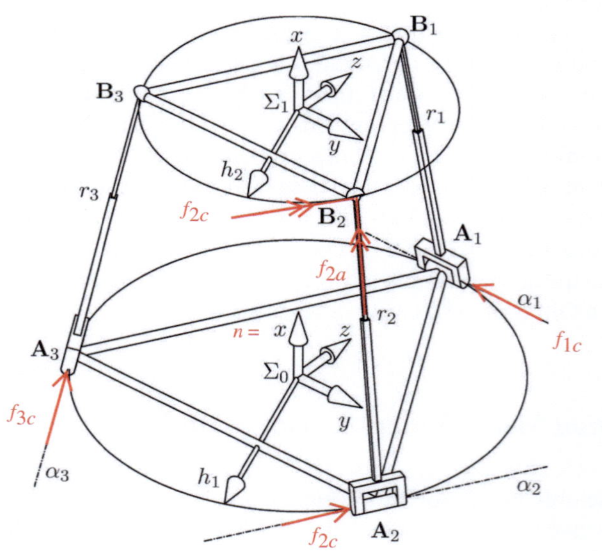

Fig. 1 3-RPS manipulator design

[1] The design parameters h_1 and h_2 (Fig. 1) have been set to $h_1 = 1$ and $h_2 = 2$. Note that all following computations can be done without specifying these parameters, but the equations become longer.

$$
\begin{aligned}
r_{00} := [&0, x_1y_3 + 2x_2x_3 + x_2y_0 - x_3y_1, -2x_1y_2 + 2x_2^2 + 2x_2y_1 - 2x_3^2 + 2x_3y_0,\\
&- R_1x_1^2 - R_1x_2^2 - R_1x_3^2 + 9x_1^2 + 12x_1y_2 + 9x_2^2 - 12x_2y_1 + x_3^2 + 4x_3y_0 + 4y_0^2 + 4y_1^2 + 4y_2^2 + 4y_3^2,\\
&(R_1 - R_2)(x_1^2 + x_2^2 + x_3^2) + 18(x_2y_1 - x_1y_2) + 6(x_3^2 - x_2^2 - x_3y_0) + 6\sqrt{3}(x_3y_1 - x_1y_3) + 2\sqrt{3}(x_2y_0 + 2x_2x_3),\\
&(R_1 - R_3)(x_1^2 + x_2^2 + x_3^2) + 18(x_2y_1 - x_1y_2) + 6(x_3^2 - x_2^2 - x_3y_0) - 6\sqrt{3}(x_3y_1 - x_1y_3) - 2\sqrt{3}(x_2y_0 + 2x_2x_3)\\
&x_1y_1 + x_2y_2 + x_3y_3, x_1^2 + x_2^2 + x_3^2 - 1],
\end{aligned} \tag{1}
$$

where R_i are the quadrances (squares) of the leg lengths r_i. Both components were characterized in Schadlbauer and Husty (2011) as follows:

- a pose of the EE frame Σ_1 in the component $x_0 = 0$ is obtained by a finite displacement of the frame Σ_0 by a π-screw (180°-screw) about a tilted axis.
- a pose of the EE frame Σ_1 in the component $x_1 = 0$ is obtained by a finite displacement of the frame Σ_0 by a screw about a horizontal axis but an arbitrary angle.

3.3 Axodes

Some authors have used the principal screw axes to describe the type of motion of a LMPM. For a 3-dof LMPM, the three principal screws meet orthogonally at a point and any screw motion of its moving platform can be expressed by a linear combination of these screws in the 3-system (Huang and Wang 2001; Bandyopadhyay and Ghosal 2004). In this paper, the axodes will be used to interpret the possible motions in each OM. The axodes of a one-parameter motion are ruled surfaces that describe the set of all instantaneous screw axes for that motion.

To obtain the axodes of possible one-parameter motions a parametrization of the motion is needed. A parametrization of the workspace of a LMPM with coupled motions is generally not possible. In case of a 3-dof LMPM one has to find three parameters that describe the same geometric object in the kinematic image space as the seven implicit equations of Eq. (1). Surprisingly this can be done for the manipulator at hand. The Eqs. (2)–(7) of r_{00} can be solved linearly for the variables R_i, y_0, y_1, y_2. The solutions for the Study parameters are:

$$
y_0 = -\frac{y_3x_1(x_1^2 + x_2^2 + x_3^2) + x_2x_3(3x_2^2 - x_3^2) + 2x_2x_3x_1^2}{x_2(x_1^2 + x_2^2 + x_3^2)}, \qquad y_1 = -\frac{x_2(x_2^2 - 3x_3^2)}{(x_1^2 + x_2^2 + x_3^2)},
$$

$$
y_2 = -\frac{-x_1x_3^3 + 3x_1x_2x_3^2 + x_3y_3x_1^2 + x_3y_3x_2^2 + x_3^3y_3}{x_2(x_1^2 + x_2^2 + x_3^2)}.
$$

The leg lengths R_i are also determined as a function of the parameters x_1, x_2, x_3, y_3. The expressions are a bit lengthy and cannot be displayed here because of lack of space. The remaining Study parameters x_1, x_2, x_3, y_3 are still linked by the eighth equation in $r_{00} : x_1^2 + x_2^2 + x_3^2 = 1$, which is simply a sphere equation in the space x_1, x_2, x_3. This sphere can be parameterized e.g. by setting: $x_2 = \cos(u), x_1 = \sin(u)\sin(v), x_3 = \sin(u)\cos(v)$. Now the whole workspace component $x_0 = 0$ is parameterized by the three parameters u, v, y_3. In matrix form the possible motions of this component can be written as:

$$\mathbf{A} = \begin{pmatrix} 1 & 0 & 0 & 0 \\ a & 1-2(\cos^2 u - \sin^2 u \cos^2 v) & 2\sin u \sin v \cos u & 2\sin^2 u \sin v \cos v \\ b & 2\sin u \sin v \cos u & -1+2\cos^2 u & 2\cos u \sin u \cos v \\ c & 2\sin^2 u \sin v \cos v & 2\cos u \sin u \cos v & 2\cos^2 v \sin^2 u - 1 \end{pmatrix}, \tag{2}$$

where

$$\begin{aligned} a &= \frac{2(-y_3 - 2\sin v \cos v \cos u + 2\sin v \cos v \cos^3 u)}{\cos u} \\ b &= -4\cos u \sin u \cos v \\ c &= 2(\cos^2 v \sin^2 u - \cos^2 u) \end{aligned} \tag{3}$$

This parametrization yields a new interpretation of the operation mode $x_0 = 0$: The manipulator is capable of all orientations determined by the two parametric set of parameters u, v. A similar interpretation is obtained for the second operation mode. If u and v take fixed values then the platform will be able to perform a translation along vector $\mathbf{n}$ normal to the base. The path of the origin in this translation is given by the vector function $(a, b, c)^T$. The motion matrix for such a one-parameter translation by setting $u = \frac{\pi}{6}$, $v = \frac{\pi}{3}$, $y_3 = t$ is given by

$$\mathbf{A}_t = \begin{pmatrix} 1 & 0 & 0 & 0 \\ -\frac{\sqrt{3}}{12}(16t+3) & -\frac{5}{8} & \frac{3}{4} & \frac{\sqrt{3}}{8} \\ -\frac{\sqrt{3}}{2} & \frac{3}{4} & \frac{1}{2} & \frac{\sqrt{3}}{4} \\ -\frac{11}{8} & \frac{\sqrt{3}}{8} & \frac{\sqrt{3}}{4} & -\frac{7}{8} \end{pmatrix}. \tag{4}$$

Having the parametrization of the workspace in Study coordinates it is relatively easy to compute the instantaneous screw axes (ISA) of any motion the manipulator can perform in one of the operation modes. Note that assembling the manipulator in one of the operation modes will only allow orientations that are given by the two parameters u, v unless the manipulator is brought into a pose where a transition into the other operation mode is possible. All poses that allow transitions are singularities. The ISA are obtained from the entries of the velocity operator:

$$\Omega = \dot{\mathbf{A}}\mathbf{A}^{-1} \tag{5}$$

To show a simple example of a one parameter motion that can be performed by the manipulator we set $u = t$, $v = \frac{1}{6}$, $y_3 = t$. This yields the motion matrix as

$$\mathbf{A}_s = \begin{pmatrix} 1 & 0 & 0 & 0 \\ \frac{-2t-\sqrt{3}\cos t+\sqrt{3}\cos^3 t}{\cos t} & -\frac{1}{2}-\frac{1}{2}\cos^2 t & \sin t\,\cos t & \frac{\sqrt{3}}{2}\sin^2 t \\ -2\sqrt{3}\cos t\,\sin t & \sin t\,\cos t & -1+2\cos^2 t & \sqrt{3}\cos t\,\sin t \\ -\frac{7}{2}\cos^2 t+\frac{3}{2} & \frac{\sqrt{3}}{2}\sin^2 t & \sqrt{3}\cos t\,\sin t & \frac{1}{2}-\frac{3}{2}\cos^2 t \end{pmatrix} \tag{6}$$

The necessary joint input parameter functions to perform the motion of Eq. 6 are already obtained from the linear solutions of the seven equations of Eq. 1. The functions are displayed in Fig. 2.

Note that completely analogous computations can be done for the second component. The equation of the ISA in the base system takes the form:

$$ISA = \begin{pmatrix} -\frac{13\sqrt{3}\cos^3 t+4t-5\sqrt{3}\cos t}{8\cos t}+\frac{\sqrt{3}}{2}\lambda \\ -\frac{-2t\,\sin t-2\cos t+13\sqrt{3}\cos^3 t\,\sin t}{4\cos^2 t} \\ -\frac{\sqrt{3}\left(13\sqrt{3}\cos^3 t+4t-5\sqrt{3}\cos t\right)}{8\cos t}+\frac{\lambda}{2} \end{pmatrix}. \tag{7}$$

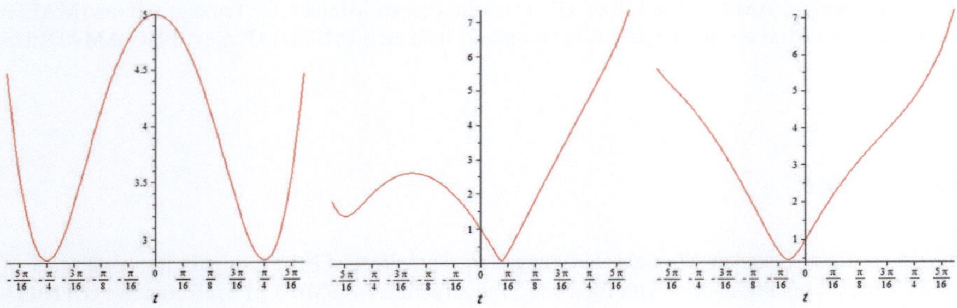

Fig. 2 Input functions for r_1, r_2, r_3

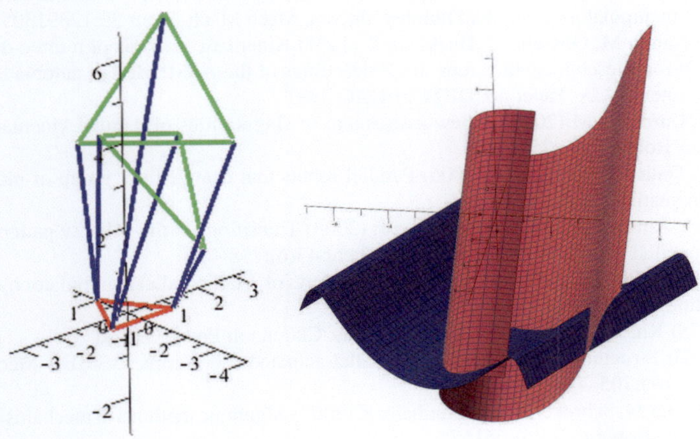

Fig. 3 Two poses of the manipulator, each in one operation mode ISA axode

This ruled surface is shown in blue in Fig. 3. The second surface is an axode of a motion in the other operation mode.

4 Conclusion

In this paper we clarified the properties of operation modes of LMPMs using the notion of Instantaneous Screw Axes (ISAs). The 3-RPS LMPM turned out to be a good illustrative example because parametrizations of both operation modes could be found. These parametrizations shed new light to the notion of 1T2R motion that was coined before for this manipulator by some authors. It also helps understand the kinematic behaviour in both modes. The axodes of one parameter motion in the operation modes were used to get a better understanding of the motion capabilities of this manipulator. It is believed that the methods developed in this paper can be of use in the analysis of other LMPMs.

Acknowledgments The authors would like to acknowledge the support of the Austrian Ministry of Science (project AMADEUS FR07/2013) and the French Ministry for Foreign Affairs (MAEE) and the French Ministry for Higher Education and Research (MESR) (Project PHC AMADEUS 2013).

References

Amine S, Tale-Masouleh M, Caro S, Wenger P, Gosselin C (2012) Singularity conditions of 3T1R parallel manipulators with identical limb structures. ASME J Mech Robot 4(1):011011–1-011011-11, 2012. doi:10.1115/1.4005336

Ball R (1900) A treatise on the theory of screws. Cambridge University Press, Cambridge

Bandyopadhyay S, Ghosal A (2004) Analytical determination of principal twists in serial parallel and hybrid manipulators using dual number algebra. Mech Mach Theor 39:12891305

Carretero J, Nahon M, Gosselin C, Buckham B (1997) Kinematic analysis of a three-dof parallel mechanism for telescope applications. In: Proceedings of the ASME design automation conference, Sacramento, CA. Paper no. DETC97/DAC-3981

Conconi M, Carricato M (2009) A new assessment of singularities of parallel kinematic chains. IEEE Trans Robot 25(4):757–770

Fanghella P, Galleti C, Gianotti E (2006) Parallel robots that change their group of motion. Adv Robot Kinematics, Springer

Hernandez A, Altuzarra O, Pinto C, Amezua E (2008) Transitions in the velocity pattern of lower mobility parallel manipulators. Mech Mach Theor 43(6):738–753

Huang Z, Wang J (2001) Identification of principal screws of 3-dof parallel manipulator by quadratic degeneration. Mech Mach Theor 36:896911

Hunt K (1978) Kinematic geometry of mechanisms. Clarendon Press, Oxford

Hunt K (1983) Structural kinematics of in-parallel-actuated robot arms. ASME J Mech Transm Autom Des 105:705–712

Husty M, Pfurner M, Schrcker H-P, Brunnthaler K (2007) Algebraic methods in mechanism analysis and synthesis. Robotica 25(6):661675

Huynh P, Hervé J (2005) Equivalent kinematic chains of three degree-of-freedom tripod mechanisms with planar spherical bonds. J Mech Des 127:95–102

Schadlbauer J, Walter DR, Husty ML (2012) A complete analysis of the 3-RPS parallel manipulator. Machines and mechanisms. Narosa Publishing House, New Delhi, India, pp 410–419, ISBN: 978-81-8487-192-0

Walter D, Husty M, Pfurner M (2009) A complete kinematic analysis of the SNU 3-UPU parallel manipulator. Contemp Math Am Math Soc 496:331–346

Walter D, Husty M (2011) Kinematic analysis of the TSAI-3UPU parallel manipulator using algebraic methods, In: Proceedings of the 13th world congress in mechanism and machine science. http://www.somim.org.mx/conference_proceedings/pdfs/A7/A7_388.pdf

Wenger P, Chablat D (1998) Workspace and assembly-modes in fully-parallel manipulators, a descriptive study, advances in robot kinematics: analysis and control. Kluwer Academic Publishers, Strobl Austria

Wohlhart K (1996) Kinematotropic linkages, recent advances in robot kinematics. In: Lenarčič J, Parenti-Castelli V (eds) Kluwer Academic, Dordrecht, pp 359–368

Zlatanov D, Bonev I, Gosselin C (2002) Constraint singularities as C-space singularities, advances in robot kinematics: theory and applications. Kluwer Academic Publishers, Dordrecht, pp 183–192

Zlatanov D, Bonev I, Gosselin C (2002) Constraint singularities of parallel mechanisms. In: Proceedings of the IEEE International conference on robotics and automation, vol 2. Washington DC, USA, pp 496–502

Geometric Contributions to the Analysis of 2-2 Wire Driven Cranes

Manfred Husty and Paul Zsombor-Murray

Abstract A 2-2 wire driven crane is the simplest wire driven manipulator. Sticking strictly to its planar motion behaviour we give some geometric insight into the possible equilibrium poses using either coupler curves of four-bar mechanisms or kinematic mapping.

Keywords Wire driven cranes · Four-bar coupler curve · Kinematic mapping

1 Introduction to the Two-Cable Planar Platform

In a recent paper Merlet (2013) Merlet states that there are only two other papers, Carricato and Merlet (2010) and Michael et al. (2009), which deal with the simplest wire driven manipulator namely the 2-2 wire driven crane. Such a two-cable planar platform is characterized by fixed supports at given points A, B connected respectively to points D, E on the platform. G is its mass centre. Triangle DEG is given as are the cable segment lengths r_A, r_B that connect A to D and B to E. The problem is to find positions of D, E, G with respect to the frame of A and B as the platform hangs in static equilibrium under the influence of a gravity force vector acting vertically downward on G, such that cables do not sustain compressive load.

In Merlet (2013) and Carricato and Merlet (2010) it is mentioned that the number of solutions to this problem is closely related to the coupler curve that is traced by the point G in the coupler motion of the coupler system DEG. But in both papers this seemingly obvious fact is not exploited geometrically. It is merely observed that a solution of the problem could be found by finding horizontal tangency points on the coupler curve. Closer inspection of the geometric properties of the coupler

M. Husty (✉)
Institute for Basic Sciences in Engineering, Unit for Geometry and CAD,
University of Innsbruck, Innsbruck, Austria
e-mail: manfred.husty@uibk.ac.at

P. Zsombor-Murray
McGill University, Montreal, Canada
e-mail: paul@cim.mcgil.ca

A. Kecskeméthy and F. Geu Flores (eds.), *Interdisciplinary Applications of Kinematics*,
Mechanisms and Machine Science 26, DOI 10.1007/978-3-319-10723-3_2

curve on the other hand will show that this is only a sufficient but not necessary condition. It is further mentioned that 12 is the maximum number of solutions to the equilibrium problem. It is not known if 12 equilibrium poses can be reached with cables not sustaining compressive load. A possible reflection of the solutions caused by a prohibited motion of the manipulator in three space will not be pursued in this paper, but it is obvious that allowing such "flip-over" will double the number of solutions.

It is believed that the two approaches to a solution to the above mentioned problem, outlined herein and all based on purely geometric principles, enjoy certain advantages compared with available solutions based on static equilibrium of forces and energy minimization. They additionally provide some geometric insight into how the analysis of spatial wire driven systems may be approached.

The paper is organized as follows. In Sect. 2 some algebraic properties of coupler curves are recalled and used for a simple proof to formally enumerate the solutions of the equilibrium problem. It will also be shown that horizontal tangency is only a sufficient condition for solution. In Sect. 3 the representation of the mechanical system in a three dimensional parameter space via kinematic mapping will be used to derive a solution method that does not depend on the horizontal tangency property of the coupler curve. Section 4 concludes the paper, in giving some ideas about solving spatial wire problems.

2 Algebraic Properties of Coupler Curves

The number of tangents from an arbitrary point in the plane to a planar algebraic curve of degree n is called the class ν of the curve. If the curve has no singularities then the class is $\nu = n(n-1)$. If the curve has simple singularities such as double points, cusps or isolated double points (acknodes) then the class is computed according to the Plücker formula Salmon (1879), p. 65 which states

$$\nu = n(n-1) - 2d - 3r, \tag{1}$$

where n is the degree of the curve, d is the number of double points and r is the number of cusps.

Using coupler curves in the solution of the 2-2 wire problem one must find horizontal tangents, which means one has to find all tangents from a point at infinity, e.g. the one that closes the horizontal x-axis, to the curve. Therefore the class of the coupler curve is needed. In Wieleitner (1908) the class of a general coupler curve is determined to be 12. This number is computed as follows: a sextic curve can have at most $\frac{1}{2}(6-1)(6-2) = 10$ double points. The coupler curve has the two circular points as triple points each with three different tangents, therefore they each count as three double points. Furthermore the coupler curve has three double points on the focal circle (see e.g. Bottema and Roth (1990), p. 338ff), therefore nine double points are known. Substituting this number into the Plücker formula

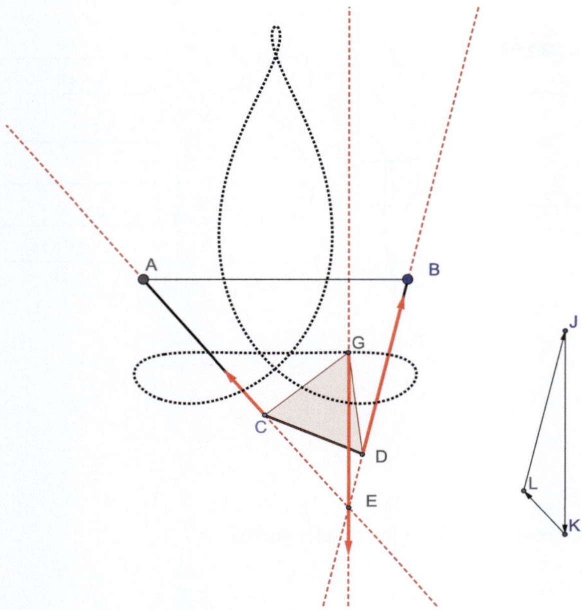

Fig. 1 Coupler curve with four double points

yields $\nu = 6(6-1) - 2 \cdot 9 = 12$. The class of the coupler curve gives therefore a simple proof that the maximum number of equilibria of a 2-2 wire system is 12. It is quite unlikely that this number can be reached with all wires in tension. Necessary conditions on the coupler curve for 12 equilibrium poses of the 2-2 wire system with all wires in tension would be:

- The coupler curve must be unicursal (it must consist of one branch). The conditions on the design parameters to fulfill this condition can be found e.g. in Bottema and Roth (1990, p. 413–418).
- The coupler curve must be on one side of the base.
- The points with horizontal tangents to the coupler curve must be between the two vertical lines passing through fixed support points A and B.

The number of coupler curve points with horizontal tangents can diminish under special design parameters. A special case arises when the four-bar is a folding four-bar. Then the coupler curve possesses a fourth double point (see Bottema and Roth (1990, p. 421)) and is rational (because it has the maximum number of simple singularities). The fourth double point is obtained by the coupler point when the four-bar is in the folded position. The Plücker formula yields $\nu = 6(6-1) - 2 \cdot 10 = 10$. Figure 1 shows the coupler curve traced by the point G with four double points. The point G is in an equilibrium position. On the right side is the force diagram. It is easily seen that this design, because of symmetry, has five equilibrium poses.

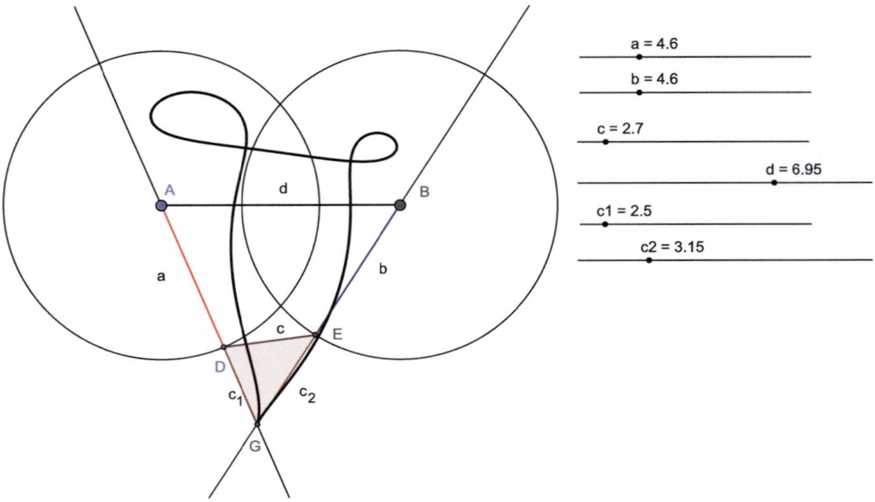

Fig. 2 Equilibrium pose at a cusp of the coupler curve

Fewer solutions arise when the coupler curve has higher order singularities or the four-bar has special design parameters (e.g. equal opposite sides of the four-bar).

Horizontal tangents to the coupler curve of at G are only a sufficient but not a necessary condition for an equilibrium pose of a 2-2-wire system. This is shown in Fig. 2. In case of a cusp in the coupler curve the two arms of the four-bar intersect at the coupler point. Therefore there are forces that balance the gravity force of G trivially.

The equation of the coupler curve is most elegantly derived using complex coordinates in the plane. Following Wunderlich (1970) the equation can be written[1]

$$[\overline{n}(z-d)P - \overline{m}zQ][n(\overline{z}-d) - m\overline{z}Q] + c^2R^2 = 0, \tag{2}$$

where

$$P = z\overline{z} + m\overline{m}c^2 - a^2, \qquad Q = (z-d)(\overline{z}-d) + n\overline{n}c^2 - b^2$$
$$R = (\overline{m} - m)z\,\overline{z} + (\overline{m}nz - m\overline{n}\,\overline{z})d. \tag{3}$$

Expanding (2), by setting $z = x + iy, \overline{z} = x - iy, m = m_1 + im_2, \overline{m} = m_1 - im_2$ ($i^2 = -1$, complex unit) yields the equation of the coupler curve $F(x, y) = 0$ in Cartesian coordinates. The points where the coupler curve has horizontal tangents are given by the intersection of the two curves

[1] This equation assumes that both base points A, B are on the x-axis. The general situation, with B not on the x-axis can be achieved by applying a simple rotation about the origin.

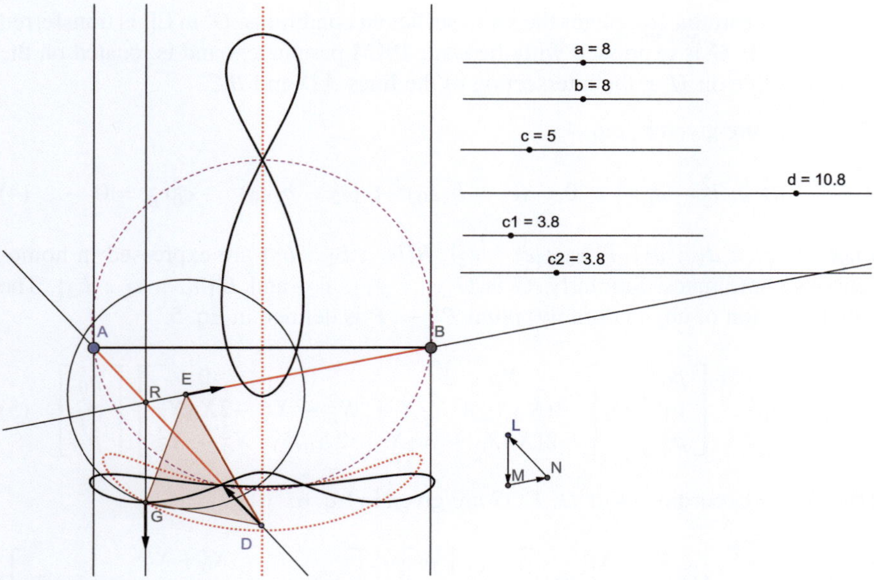

Fig. 3 Equilibrium pose with curve $\frac{\partial F}{\partial x} = 0$

$$F(x, y) = 0, \qquad \frac{\partial F}{\partial x} = 0.$$

Figure 3 shows a two wire system in a equilibrium pose. The red dotted curve represents $\frac{\partial F}{\partial x} = 0$. G is at an intersection point of both curves. The plot shows that this system has six real equilibrium poses with positive forces in the wires.[2] Note that also the double points are on the intersection of both curves. The double points can be easily found because they are, as mentioned above, also on the focal circle. The focal circle is given by the equation $R = 0$ (Eq. 3).

3 Kinematic Mapping

A solution based on planar kinematic mapping (PKM) solves for the positions of D, E, G directly. The notion introduced above is used. It proceeds as follows.

- Point D', expressed in the moving or end effector frame EE, is mapped to D on the circle of radius $r_A = a$ and centred on A expressed in the fixed or base frame FF.
- Similarly, E' is mapped to E on circle, radius $r_B = b$, centred on B.

[2] Numerically stable curve intersection computation can be achieved. A visualization can help to find good starting values for a Newton method.

- A third constraint completes the set of sufficient conditions. G' in EE is transferred to G in FF. G is expressed with the same PKM parameters and is located on the vertical line on H at the intersection of the lines AD and BE.

The circles are given by Eq. 4

$$d_1^2 + d_2^2 - d_0^2 r_A^2 = 0, \quad (e_1 - b_1 e_0)^2 + (e_2 - b_2 e_0)^2 - e_0^2 r_B^2 = 0 \tag{4}$$

where $D\{d_0 : d_1 : d_2\}$, $E\{e_0 : e_1 : e_2\}$, $B\{b_0 : b_1 : b_2\}$ are expressed in homogeneous coordinates. Similarly, G is $G\{g_0 : g_1 : g_2\}$ and $H\{h_0 : h_1 : h_2\}$. The transformation of any (a dummy) point $P' \to P$ is defined in Eq. 5.

$$\mathbf{p} = [\mathbf{X}]\mathbf{p}' : \begin{bmatrix} p_0 \\ p_1 \\ p_2 \end{bmatrix} = \begin{bmatrix} X_0^2 + X_3^2 & 0 & 0 \\ 2(X_0X_2 + X_1X_3) & X_0^2 - X_3^2 & -2X_0X_3 \\ -2(X_0X_1 - X_2X_3) & 2X_0X_3 & X_0^2 - X_3^2 \end{bmatrix} \begin{bmatrix} p_0' \\ p_1' \\ p_2' \end{bmatrix} \tag{5}$$

Transformed coordinates of D, E, G are given in Eq. 6.

$$\mathbf{d} = \begin{bmatrix} d_0 \\ d_1 \\ d_2 \end{bmatrix} = \begin{bmatrix} X_0^2 + X_3^2 \\ 2(X_0X_2 + X_1X_3) \\ -2(X_0X_1 - X_2X_3) \end{bmatrix}, \mathbf{e} = \begin{bmatrix} e_0 \\ e_1 \\ e_2 \end{bmatrix} = \begin{bmatrix} X_0^2 + X_3^2 \\ 2(X_0X_2 + X_1X_3) + e_1'(X_0^2 - X_3^2) \\ -2(X_0X_1 - e_1'X_0X_3 - X_2X_3) \end{bmatrix},$$

$$\mathbf{g} = \begin{bmatrix} g_0 \\ g_1 \\ g_2 \end{bmatrix} = \begin{bmatrix} X_0^2 + X_3^2 \\ 2(X_0X_2 - g_2'X_0X_3 + X_1X_3) + g_1'(X_0^2 - X_3^2) \\ -2(X_0X_1 - g_1'X_0X_3 - X_2X_3) + g_2'(X_0^2 - X_3^2) \end{bmatrix} \tag{6}$$

Before writing the three constraint equations in mapping parameters consider the line segments $AD \equiv a\{A_0 : A_1 : A_2\}$, $BE \equiv b\{B_0 : B_1 : B_2\}$, their intersection $H = a \cap b$ and the vertical line $GH \equiv g\{G_0 : G_1 : G_2\}$.

$$\begin{vmatrix} p_0 & p_1 & p_2 \\ 1 & a_1 & a_2 \\ d_0 & d_1 & d_2 \end{vmatrix} = A_0p_0 + A_1p_1 + A_2p_2 = (a_1d_2 - a_2d_1)p_0 + (a_2d_0 - d_2)p_1 + (d_1 - a_1d_0)p_2 = 0 \tag{7}$$

$$\begin{vmatrix} p_0 & p_1 & p_2 \\ 1 & b_1 & b_2 \\ e_0 & e_1 & e_2 \end{vmatrix} = B_0p_0 + B_1p_1 + B_2p_2 = (b_1e_2 - b_2e_1)p_0 + (b_2e_0 - e_2)p_1 + (e_1 - b_1e_0)p_2 = 0 \tag{8}$$

$$\begin{vmatrix} P_0 & P_1 & P_2 \\ A_0 & A_1 & A_2 \\ B_0 & B_1 & B_2 \end{vmatrix} = h_0P_0 + h_1P_1 + h_2P_2 = (A_1B_2 - A_2B_1)P_0 + (A_2B_0 - A_0B_2)P_1 + (A_0B_1 - A_1B_0)P_2 = 0 \tag{9}$$

$$\begin{vmatrix} p_0 & p_1 & p_2 \\ g_0 & g_1 & g_2 \\ h_0 & h_1 & h_2 \end{vmatrix} = G_0p_0 + G_1p_1 + G_2p_2 = (g_1h_2 - g_2h_1)p_0 + (g_2h_0 - g_0h_2)p_1 + (g_0h_1 - g_1h_0)p_2 = 0 \tag{10}$$

Since g is a vertical line the third constraint equation, expressing the redundant situation that these lines intersect on a common point, i.e., $a \cap b \cap g = H$, reduces to Eq. 11. The coefficient of the y-coordinate in an equation of a vertical line must vanish.

$$G_2 = g_0 h_1 - g_1 h_0 = g_0(b_1 e_2 - b_2 e_1)d_1 - g_1[(b_1 e_0 - e_1)d_2 - (b_2 e_0 - e_2)d_1] = 0 \quad (11)$$

Substituting the results from the first two of Eqs. 6 respectively into the pair Eq. 4, dehomogenizing by setting $X_0 = 1$, subtracting the first circle equation from the second, removing a common factor $2(X_0^2 + X_3^2)$ and normalizing on the length interval between D and E so that $e_1' = 1$ yields two quadric constraints, Eq. 12,

$$4(X_1^2+X_2^2)-r_A^2(X_3^2+1) = 0, \quad j_1+j_2X_1+j_3X_2-j_2X_3+j_4X_1X_3-j_2X_2X_3+j_5X_3^2 = 0 \quad (12)$$

where

$$j_1 = (b_1-1)^2 + b_2^2 + r_A^2 - r_B^2, \quad j_2 = 4b_2, \quad j_3 = -4(b_1-1),$$
$$j_4 = -4(b_1+1), \quad j_5 = (b_1+1)^2 + b_2^2 + r_A^2 - r_B^2 \quad (13)$$

Incorporating all the choices of origins, direction and length into Eqs. 7 through 10 and combining this with the first two of Eq. 6 to get h_0, h_1 then using the third of Eq. 6 to get g_0, g_1, directly, reveals Eq. 11 as the quartic, Eq. 14,

$$k_1X_1 + k_2X_2 - 2X_1X_2 + k_3X_1X_3 + k_4X_2X_3 - 2X_1^2X_3 - 2X_2^2X_3$$
$$+k_5X_1X_3^2 + k_6X_2X_3^2 - 2X_1X_2X_3^2 + k_7X_1X_3^3 + k_8X_2X_3^3 = 0 \quad (14)$$

where

$$k_1 = (b_1-1)g_1', \quad k_2 = (g_1'-1)b_2, \quad k_3 = k_2 + 2(1-b_1)g_2', \quad k_4 = 2(b_1 - b_2g_2') - k_8$$
$$k_5 = 2(b_1 - b_2g_2') - k_1, \quad k_6 = 2(b_1+1)g_2' - k_2, \quad k_7 = -k_2, \quad k_8 = (1+b_1)g_1'$$

The two quadric constraints, Eq. 12, and the quartic, Eq. 14, indicate that a univariate of degree no greater than 16 in, say, X_3 should be available as a solution to the planar problem. Actually, one may do better. Elimination of X_1 from the two equations Eq. 12 and from the second of that pair and Eq. 14 produces a bivariate pair of equations whose resultant with respect to X_2 yields a univariate of degree 12 in X_3 and the factor $[8\sqrt{2}\{(b_1+1)X_3 - b_2\}]^4$. This elimination can be done entirely with symbolic design parameters but unfortunately not in their compressed form as given by $j_i,\ i = 1, \ldots, 5$ and $k_j,\ j = 1, \ldots, 8$ in Eqs. 12 and 14. To effect backsubstitution so that X_1, X_2 can be calculated uniquely for each of (up to) 12 real values of X_3, X_1 is removed between the two Eqs. 12 and between the left hand equation of Eqs. 12 and 14. These two equations, devoid of X_1, are quadratic in X_2 so the coefficients of X_2^2 in each can be used as multipliers in the other equation and the difference will be linear in X_2. With values of X_2, X_3 available, Eq. 14, which is linear in both X_1 and X_2, will furnish X_1. A numerical example with six real solutions, all of them with compression-free legs, was computed, without loss in generality, with the same design parameters as in Fig. 3. The results, are tabulated below. Their disposition shown in Fig. 4.

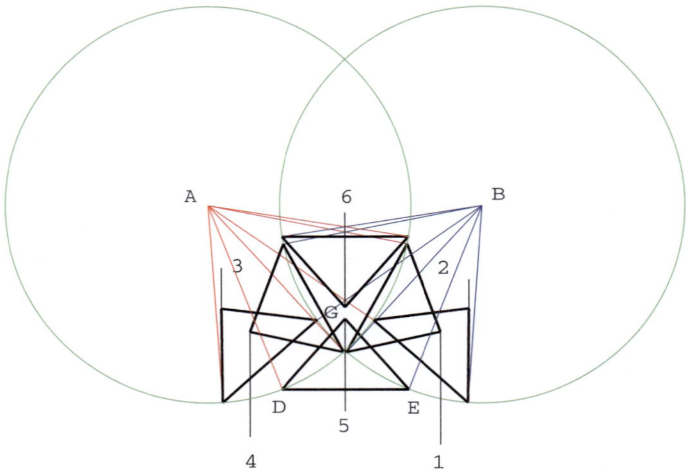

Fig. 4 Six feasible poses

Table 1 Numerical example with six real solutions

Solution	D	E	G
1	(1.5693, −0.31205)	(1.0906, −1.1901)	(1.8325, −1.0250)
2	(1.3024, −0.92906)	(2.0476, −1.5958)	(2.0566, −0.83590)
3	(0.11229, −1.5961)	(0.85751, −0.92923)	(0.10325, −0.83611)
4	(1.0694, −1.1901)	(0.59076, −0.31210)	(0.32754, −1.0251)
5	(-0.5800, −1.4912)	(1.5797, −1.4912)	(1.0802, −2.0634)
6	(1.5797, −0.2522)	(0.5800, −0.2522)	(1.0802, −0.8244)

4 Conclusion

PKM solves the platform pose problem directly without resort to a four-bar coupler curve and its positions of zero slope. The univariate polynomial can be computed symbolically in terms of design parameters but its coefficients are too large to be practical unless a more efficient means of compression can be found. Using a formulation in complex number notation provides a way to examine the coupler curve. Its positions of zero slope can be found by expanding the curve equation to Cartesian coordinates and intersection with the curve's derivative with respect to the x coordinate. The univariate can be computed completely symbolically and reveals apparent solutions at cusps at non-zero slope. It is proposed to extend this work using *spatial* kinematic mapping with a three sphere paradigm. Three additional constraints must be found. The three rank deficient determinants provided by the vertical fourth line define necessary but insufficient conditions (Table 1).

Acknowledgments This research is supported by a Natural Sciences & Engineering Canada (NSERC) "Discovery" research grant.

References

Bottema O, Roth B (1990) Theoretical kinematics. Dover Publications, New York

Carricato M, Merlet J.-P (2010) Geometrico-static analysis of under-constrained cable-driven parallel robots. In: Lenarčič J, Stanisič M (eds) Advances in robot kinematics. Springer, pp 309–320

Gfrerrer A (2012) Static equilibrium of a rigid body in the context of line geometry, private communication 12–12-21, 1 p

Merlet J.-P (2013) Further analysis of the 2–2 wire driven parallel crane, presented at CK 2013, p 8

Michael N, Fink J, Kumar V (2009) Cooperative manipulation and transportation with aerial robots. In: Robotics: science and systems, Seattle

Salmon G (1879) A treatise on higher planar curves, 3rd ed. Dublin

Wieleitner H (1908) Spezielle ebene Kurven. Göschen Verlagshandlung, Leipzig

Wunderlich W (1970) Ebene Kinematik v. 447/447a, B.I. Hochschultaschenbücher, Bibiliographisches Institut, Mannheim, pp 66–69

Zsombor-Murray PJ, Gfrerrer A (2010) A unified approach to the direct kinematics of some reduced motion parallel manipulators. ASME J Mech Rob 2(2):10. doi:10.1115/1.4001095

Zsombor-Murray PJ (2006) Planar kinematic mapping fundamentals. http://www.cim.mcgill.ca/~paul/PKMF6Ac.pdf

Design of Cable-Driven Parallel Robots with Multiple Platforms and Endless Rotating Axes

Philipp Miermeister and Andreas Pott

Abstract Cable-driven parallel robots allow high performance operation due to their minimal actuated system mass. In order to deal with more complex handling operations it is desirable to have an additional actuated serial kinematics on the platform. This usually comes along with the problem of extra weight and energy supply for the motors on the platform. In this paper we present a new approach to the problem by introducing a hybrid design with coupled platforms and cable-driven serial kinematics. Especially the concept of an endless rotatable axis will be highlighted.

Keywords Cable-driven parallel robots · Kinematics · Robot design · Multiple platforms

1 Introduction

Cable-driven parallel robots (hereafter termed cable robots) mainly consist of a light weight platform, cables, and winches and therefore allow to design systems with an exceptional good power to mass ratio compared to conventional kinematics. The workspace of the robot mainly depends on the winch positions and platform attachment points, but in general cable robots have a relatively small rotational workspace compared to their translational workspace. Previous works related to robot design and workspace analysis can be found in Gouttefarde et al. (2007), Ming and Higuchi (1994), Merlet and Daney (2007). Some use-cases for cable robots as handling operations may demand a larger rotational workspace which can be achieved by additional actuated axes on the platform. Additional motors on the platform add weight to the system and therefore reduce the performance. Furthermore, one has to deal with the problem of signal transmission and power supply.

P. Miermeister (✉) · A. Pott
Fraunhofer IPA, Nobelstraße 12, 70569 Stuttgart, Germany
e-mail: philipp.miermeister@ipa.fraunhofer.de

A. Pott
e-mail: andreas.pott@ipa.fraunhofer.de

A. Kecskeméthy and F. Geu Flores (eds.), *Interdisciplinary Applications of Kinematics*, Mechanisms and Machine Science 26, DOI 10.1007/978-3-319-10723-3_3

In this paper a new approach is presented were the additional axes are actuated by the platform cables. This keeps the mass of the system low and avoids the need for power supply. Beside that, additional cables increase the safety of the system and may even be used to increase the translational workspace. Related work to cable-driven robots with hybrid kinematic structures can be found in Kossowski and Notash (2002), Varziri and Notash (2007).

In the following, the basic kinematics and force equations are derived and examples for cable-driven axes are shown. The first concept shows the most simple realization of a single rigid cable-driven platform with a cable-driven axis mounted on it. The second concept deals with a series of separately actuated platform parts which are connected using elementary joints and a single rigid transmission element. Controlling the relative position of the platform parts causes the axis to move in the desired direction.

2 Platform Kinematics for Single Platform

Cable-driven parallel robots consist of a single platform which is connected to m winches on a frame using a set of cables. The platform position $\mathbf{r}$ and rotation $\mathbf{R}$ can be controlled by changing the cable length $\mathbf{l_i}$ according to the inverse kinematics

$$\mathbf{l}_i = \mathbf{a}_i - \mathbf{r} - \mathbf{R}\mathbf{b}_i , \tag{1}$$

where the vectors $\mathbf{a}_i$ relate to the cable's outlet points at the winch side and $\mathbf{b_i}$ are the distal anchor points on the mobile platform. Considering the platform as a free floating body, a stable platform position is characterized by the force and torque equilibrium

$$\mathbf{A}^{\mathrm{T}}\mathbf{f} + \mathbf{w} = \mathbf{0}, \tag{2}$$

where $\mathbf{f}$ and $\mathbf{w}$ denote the cable forces and external wrench respectively while $\mathbf{A}^{\mathrm{T}}$ relates to the well known structure matrix Verhoeven (2004)

$$\mathbf{A}^{\mathrm{T}} = \begin{bmatrix} \mathbf{u}_1 & \cdots & \mathbf{u}_m \\ \mathbf{b}_1 \times \mathbf{u}_1 & \cdots & \mathbf{b}_m \times \mathbf{u}_m \end{bmatrix}. \tag{3}$$

The unit vectors $\mathbf{u}_i$ describe the direction of the cables. Cables can only resist pulling forces so that Eq. (2) must be fulfilled under the constrained of positive forces $f_i > 0$ for $i = 1..m$. For the sake of simplicity we limit the considerations to wrench closure where in practice a minimum and maximum force must be regarded.

3 System of Connected Platforms

In the following we extend the modeling of cable robots to multiple platforms. For simplicity, only systems of platforms with a serial topology as shown in Fig. 1 are considered. The whole system consists of k bodies. Rigid bodies in the system which are connected to cables are referred to as platform. The number of platforms is denoted by y. Elements which are not connected to cables are assumed to be fully constrained by joints and will be referred to as transmission elements whose number is $t = k - y$. The total number of cables of all platforms together is labeled by m. Each of the y platforms has a subset of m_i with $\sum m_i = m$ cables attached to it. The whole assembled system of platforms and joints without cables has $n = \sum_{i=1}^{k} n_i - n_c$ degree of freedom (DOF) where n_i is the DOF of each body and n_c summarizes all constraints in the system. For a free floating mechanism the DOF of each platform is $n_i = 6$. For single platform it is necessary to have at least $m = n + 1$ cables since cables can only resist pulling forces. Therefore a system of y connected platforms and n_c constraints needs at least $m = n + y$ cables. Each single platform i in the system with n_{ci} constraints has to be connected to at least

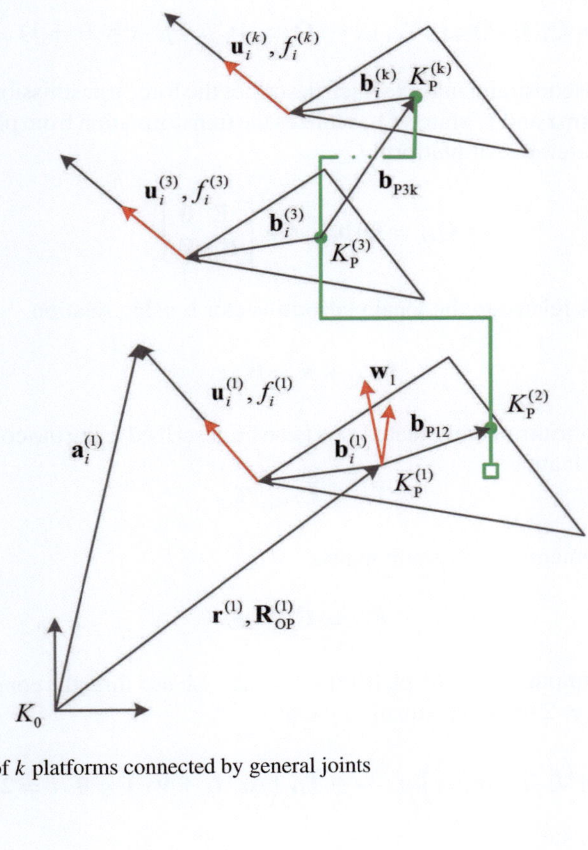

Fig. 1 Series of k platforms connected by general joints

$$m_i \geq n_i - n_{ci} + 1 \tag{4}$$

cables. For the following definitions, the first platform $i = 1$ is regarded as reference platform and the kinematic chain is labeled with increasing numbers, where the actuated axis usually has index $i = 2$. The structure equation for the first platform is given by

$$\mathbf{A}_1{}^{\mathrm{T}}\mathbf{f}_1 + \mathbf{w}_1 + \mathbf{w}_{\mathrm{p},12} + \cdots + \mathbf{w}_{\mathrm{p},1\mathrm{k}} = \mathbf{0} \tag{5}$$

where $\mathbf{w}_{\mathrm{p,i}}$ is the wrench of the i-th platform and $\mathbf{w}_1$ is the external wrench of the platform itself containing gravity and process forces. Introducing the transformation matrix $\mathbf{T}_{ij}$ which is used to describe the force transmission between platform i and j, the applied wrenches are obtained from

$$\mathbf{w}_{\mathrm{p},ij} = \mathbf{T}_{ij}\left(\mathbf{A}_j{}^{\mathrm{T}}\mathbf{f}_j + \mathbf{w}_j\right). \tag{6}$$

The transformation matrix $\mathbf{T}_{ij}$ includes the joint constraints and geometric relations and can be computed by

$$\mathbf{T}_{ij} = \mathbf{Q}_{is}\mathbf{C}_{is}\mathbf{Q}_{i+1,s}\mathbf{C}_{i+1,s}\cdots\mathbf{Q}_{j-1,j}\mathbf{C}_{j-1,j} \quad s = (i+1)\ldots j \tag{7}$$

where $\mathbf{C}_{ij}$ is the constraint matrix which describes the force transmission of the joints between platform i and j. Matrix $\mathbf{Q}_{ij}$ realizes the transformation from platform frame j to the local reference of platform i

$$\mathbf{Q}_{ij} = \mathbf{Q}(\mathbf{b}_{\mathrm{p},ij}) = \begin{bmatrix} \mathbf{E} & \mathbf{0} \\ \tilde{\mathbf{B}}_{ij} & \mathbf{E} \end{bmatrix}. \tag{8}$$

The matrix $\mathbf{B}$ is related to the local platform vector $\mathbf{b}$ using relation

$$\mathbf{b}_{\mathrm{P},ij} \times \mathbf{v} = \tilde{\mathbf{B}}_{ij}\mathbf{v}. \tag{9}$$

The force equilibrium of the actuated axis i can be described using the complementary transformation matrix

$$\overline{\mathbf{T}}_{ij} = \overline{\mathbf{C}}_{i-1,i}\mathbf{T}_{ij} \tag{10}$$

and the complementary constraint matrix

$$\bar{\mathbf{C}}_{ij} = \mathbf{C}_{ij} - \mathbf{E}. \tag{11}$$

Under the assumption that all platforms $i = 2\ldots k$ are directly connected to the actuated axis $i = 2$ the force equation yields

$$\overline{\mathbf{T}}_{i,i+1}\left(\mathbf{A}_{i+1}{}^{\mathrm{T}}\mathbf{f}_{i+1} + \mathbf{w}_{i+1}\right) + \cdots + \overline{\mathbf{T}}_{ik}\left(\mathbf{A}_k{}^{\mathrm{T}}\mathbf{f}_k + \mathbf{w}_k\right) = \mathbf{0} \quad i = 2\ldots k. \tag{12}$$

The whole system of platforms, applied cable forces, and wrenches now can be written as

$$\begin{bmatrix} \mathbf{A}_1^T & \mathbf{T}_{13}\mathbf{A}_3^T & \cdots & \mathbf{T}_{1k}\mathbf{A}_k^T \\ \mathbf{0} & \overline{\mathbf{T}}_{23}\mathbf{A}_3^T & \cdots & \overline{\mathbf{T}}_{2k}\mathbf{A}_k^T \\ \mathbf{0} & \overline{\mathbf{C}}_{23}\mathbf{A}_3^T & \mathbf{0}\cdots & \mathbf{0} \\ \mathbf{0} & \mathbf{0} & \ddots & \mathbf{0} \\ \mathbf{0} & \mathbf{0} & \mathbf{0} & \overline{\mathbf{C}}_{2k}\mathbf{A}_k^T \end{bmatrix} \begin{bmatrix} \mathbf{f}_1 \\ \mathbf{f}_3 \\ \vdots \\ \mathbf{f}_k \end{bmatrix} + \begin{bmatrix} \mathbf{E} & \mathbf{T}_{12} & \mathbf{T}_{13} & \cdots & \mathbf{T}_{1k} \\ \mathbf{0} & \overline{\mathbf{C}}_{12} & \overline{\mathbf{T}}_{23} & \cdots & \overline{\mathbf{T}}_{2k} \\ \mathbf{0} & \mathbf{0} & \overline{\mathbf{C}}_{23} & \mathbf{0}\cdots & \mathbf{0} \\ \mathbf{0} & \mathbf{0} & \mathbf{0} & \ddots & \mathbf{0} \\ \mathbf{0} & \mathbf{0} & \mathbf{0} & \mathbf{0} & \overline{\mathbf{C}}_{2k} \end{bmatrix} \begin{bmatrix} \mathbf{w}_1 \\ \mathbf{w}_2 \\ \mathbf{w}_3 \\ \vdots \\ \mathbf{w}_k \end{bmatrix} = \mathbf{0}. \quad (13)$$

Here we showed that multi-platform robots can be modeled with the same structure and class of equations as the known structure equation $\mathbf{A}^T\mathbf{f} + \mathbf{w} = \mathbf{0}$ for single-platform robots. Therefore we can use existing algorithms for their design and analysis.

3.1 Designs for an Endless Rotating Axis

A simple design for a cable robot with a cable-driven axis is shown in Fig. 2a, b. This platform consists of $k = y = 2$ bodys which are connected by a revolute joint with $n_c = 5$. This setup results in a system with $n = kn_i - n_c = 7$ DOF and therefore at least $m = 8$ cables are needed. The axis in direction of the normalized vector $\mathbf{n}_a$ is supported by a revolute joint whose location is described by vector $\mathbf{b}_r$. Actuation of the axis is provided by m_p cables that are fixed at a single pivot point at the end of a cantilever denoted by $\mathbf{b}_c$. The platform is supported by m_a cables. With the appropriate geometry parameters, this configuration allows infinite rotation of the axis by moving the pivot point along a circular trajectory. The actuation torque of the axis caused by the cables is computed using the orthogonal projection matrix $\overline{\mathbf{C}}_T = \mathbf{n}_a\mathbf{n}_a^T$ so that the torque equilibrium yields $\overline{\mathbf{C}}_T \left(\left[\, \mathbf{b}_p(\phi) \times \mathbf{u}_{a,1} \cdots \mathbf{b}_p(\phi) \times \mathbf{u}_{a,m_a} \right] \mathbf{f}_a\right) + \overline{\mathbf{C}}_T \tau_a = \mathbf{0}$ where ϕ describes the set point of the endeffector. Since the revolute joint constrains

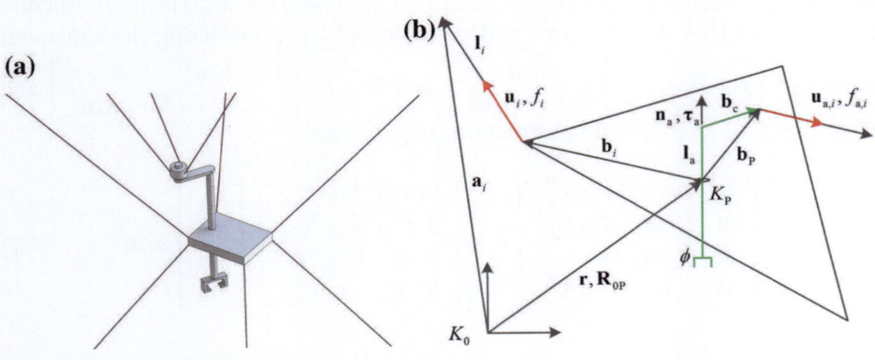

Fig. 2 Platform design with cable-driven endless rotating axis

all translational movements and absorbs forces in all direction it follows $\overline{\mathbf{C}}_\mathrm{F} = \mathbf{0}$ and $\overline{\mathbf{C}}_2 = \begin{bmatrix} \mathbf{0} & \mathbf{0} \\ \mathbf{0} & \overline{\mathbf{C}}_\mathrm{T} \end{bmatrix}$. Under consideration of the pivot point $\mathbf{b}_\mathrm{p}(\phi) = \mathbf{b}_\mathrm{r} + \mathbf{l}_\mathrm{a} + \mathbf{R}_\mathrm{z}(\phi)\mathbf{b}_\mathrm{c}$ one can compute the force and torque which is acting on the platform P by

$$\mathbf{w}_\mathrm{P} = \underbrace{\begin{bmatrix} \mathbf{E} & \mathbf{0} \\ \mathbf{0} & (\mathbf{E} - \overline{\mathbf{C}}_\mathrm{T}) \end{bmatrix}}_{\mathbf{C}_2} \mathbf{w}_{\mathrm{P}'} \tag{14}$$

where

$$\mathbf{w}_{\mathrm{P}'} = \underbrace{\begin{bmatrix} \mathbf{u}_{\mathrm{a},1} & \cdots & \mathbf{u}_{\mathrm{a,m}} \\ \mathbf{b}_\mathrm{p} \times \mathbf{u}_{\mathrm{a},1} & & \mathbf{b}_\mathrm{p} \times \mathbf{u}_{\mathrm{a,m}} \end{bmatrix}}_{\mathbf{A}_2^\mathrm{T}} \begin{bmatrix} f_{\mathrm{a},1} \\ \vdots \\ f_{\mathrm{a},m} \end{bmatrix} + \mathbf{w}_\mathrm{a} \tag{15}$$

with $\mathbf{w}_\mathrm{a} = \begin{bmatrix} \mathbf{f}_\mathrm{a} & \tau_\mathrm{a} \end{bmatrix}^\mathrm{T}$. Structure matrix $\mathbf{A}_2^\mathrm{T}$ depends on rotation ϕ of the axis. The structure equation of the whole system reads

$$\begin{bmatrix} \mathbf{A}_1^\mathrm{T} & \mathbf{C}_2\mathbf{A}_2^\mathrm{T} \\ \mathbf{0} & \overline{\mathbf{C}}_2\mathbf{A}_2^\mathrm{T} \end{bmatrix} \mathbf{f} + \begin{bmatrix} \mathbf{E} & \mathbf{C}_2 \\ \mathbf{0} & \overline{\mathbf{C}}_2 \end{bmatrix} \begin{bmatrix} \mathbf{w}_1 \\ \mathbf{w}_2 \end{bmatrix} = \mathbf{0}. \tag{16}$$

For a robot design with an actuated linear axis instead of the rotary axis, the associated linear joint would be described by $\mathbf{C}_2 = \begin{bmatrix} \mathbf{C}_\mathrm{F} & \mathbf{0} \\ \mathbf{0} & \mathbf{0} \end{bmatrix}$ with $\mathbf{C}_\mathrm{F} = \mathbf{n}_\mathrm{a}\mathbf{n}_\mathrm{a}^\mathrm{T}$. A more complex robot design with $n = 4$ bodys or respectively $y = 3$ cable-driven platforms is shown in Fig. 3 whose kinematic structure is shown in Fig. 1. In the model, all three joints have one rotational degree of freedom in direction of the vertical axis and the joint between body 2, 3 has an additional translational degree of freedom in direction of the vertical axis which for example is used to open and close the gripper. The minimal number of cables can be compute from $n = 24$ and $n_c = 14$ with $n - n_c + y = 13$. The number of cables attached to each platform is given by $m_1 = 6, m_2 = 0, m_3 = 4$, and $m_4 = 3$ fulfilling condition Eq. (4). The local structure matrix for each of the platforms is denoted by $\mathbf{A}_{1,3,4}^\mathrm{T}$. Considering the constraint matrices $\mathbf{C}_{12} = \mathbf{C}_{24} = \begin{bmatrix} \mathbf{E} & \mathbf{0} \\ \mathbf{0} & \mathbf{E} - \mathbf{n}_\mathrm{a}\mathbf{n}_\mathrm{a}^\mathrm{T} \end{bmatrix}$ and $\mathbf{C}_{23} = \begin{bmatrix} \mathbf{E} - \mathbf{n}_\mathrm{a}\mathbf{n}_\mathrm{a}^\mathrm{T} & \mathbf{0} \\ \mathbf{0} & \mathbf{E} - \mathbf{n}_\mathrm{a}\mathbf{n}_\mathrm{a}^\mathrm{T} \end{bmatrix}$ and using Eqs. (7–12), the resulting structure equation reads

$$\begin{bmatrix} \mathbf{A}_1^\mathrm{T} & \mathbf{T}_{13}\mathbf{A}_3^\mathrm{T} & \mathbf{T}_{14}\mathbf{A}_4^\mathrm{T} \\ \mathbf{0} & \overline{\mathbf{T}}_{23}\mathbf{A}_3^\mathrm{T} & \overline{\mathbf{T}}_{24}\mathbf{A}_4^\mathrm{T} \\ \mathbf{0} & \overline{\mathbf{C}}_{23}\mathbf{A}_3^\mathrm{T} & \mathbf{0} \\ \mathbf{0} & \mathbf{0} & \overline{\mathbf{C}}_{24}\mathbf{A}_{44}^\mathrm{T} \end{bmatrix} \mathbf{f} + \begin{bmatrix} \mathbf{E} & \mathbf{T}_{12} & \mathbf{T}_{13} & \mathbf{T}_{14} \\ \mathbf{0} & \overline{\mathbf{C}}_{12} & \overline{\mathbf{T}}_{23} & \overline{\mathbf{T}}_{24} \\ \mathbf{0} & \mathbf{0} & \overline{\mathbf{C}}_{23} & \mathbf{0} \\ \mathbf{0} & \mathbf{0} & \mathbf{0} & \overline{\mathbf{C}}_{24} \end{bmatrix} \begin{bmatrix} \mathbf{w}_1 \\ \mathbf{w}_2 \\ \mathbf{w}_3 \\ \mathbf{w}_4 \end{bmatrix} = \mathbf{0}. \tag{17}$$

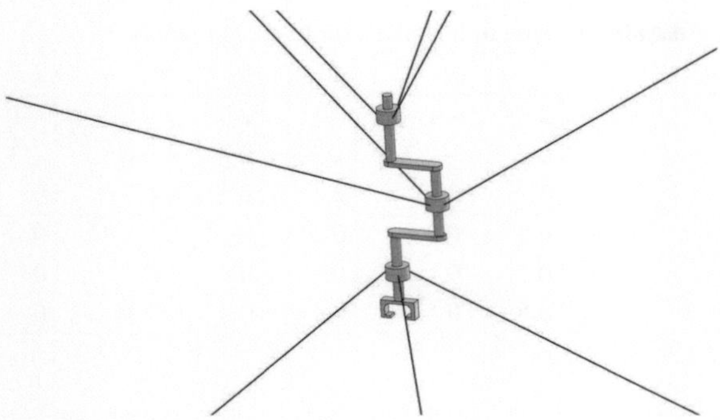

Fig. 3 3D model of a cable robot with three platforms

Wrench $\mathbf{w}_{1,4}$ only contains the gravity force of the respective platform while $\mathbf{w}_2$ additionally includes the applied process forces and $\mathbf{w}_3$ includes the vertical force caused by the gripper mechanics.

3.2 Simulation Results

In the following we present an example of a cable-driven parallel robot with an endless rotation around its z-axis. The design of the robot is based on a simple 3–9 design with a triangular frame structure and a planar mobile platform similar to the geometric structure shown in Fig. 2b. The geometry data of the sample robot can be found in Table 1. Now we consider a trajectory that includes a full rotation around the z-axis of the mobile platform. Let $\mathbf{r} = [0, 0.5, 1.5]^T$ be the position of the platform. Furthermore, the orientation of the platform $\mathbf{R}$ is choosen to be the elementary rotation matrix $\mathbf{R}_z(\phi)$. The force distributions was computed using the closed-form method Pott et al. (2009). In Fig. 4 the computation results for all nine cables are shown for $\phi \in [0, 2\pi]$. It can be seen that the value of all nine forces is continuous and between the force bounds $f_{\text{min}} = 1$ and $f_{\text{max}} = 10$. Therefore, the platform is capable to perform a full rotation around the z-axis with possible endless repetition. The example shows the prove of concept for a robot design with unlimited rotation capacities for at least one of its axis. Although it cannot be seen from the diagram, the cables do not intersect at any time. According to Verhoeven (2004) two cables can only intersect at one point. Since three sets of each three cables share a common anchor point, these cables cannot intersect during the motion.

Table 1 Geometry data for the base $\mathbf{a}_i$ and platform $\mathbf{b}_i$ anchor points

	i	1	2	3	4	5	6	7	8	9
Base $\mathbf{a}_i$	x	-1	2	-1	-1	2	-1	-1	2	-1
	y	$\sqrt{3}$	0	$-\sqrt{3}$	$\sqrt{3}$	0	$-\sqrt{3}$	$\sqrt{3}$	0	$-\sqrt{3}$
	z	3	3	3	2	2	2	0	0	0
Platform $\mathbf{b}_i$	x	0.3	0.3	0.3	0	0	0	0	0	0
	y	0	0	0	0	0	0	0	0	0
	z	0.5	0.5	0.5	0.4	0.4	0.4	0	0	0

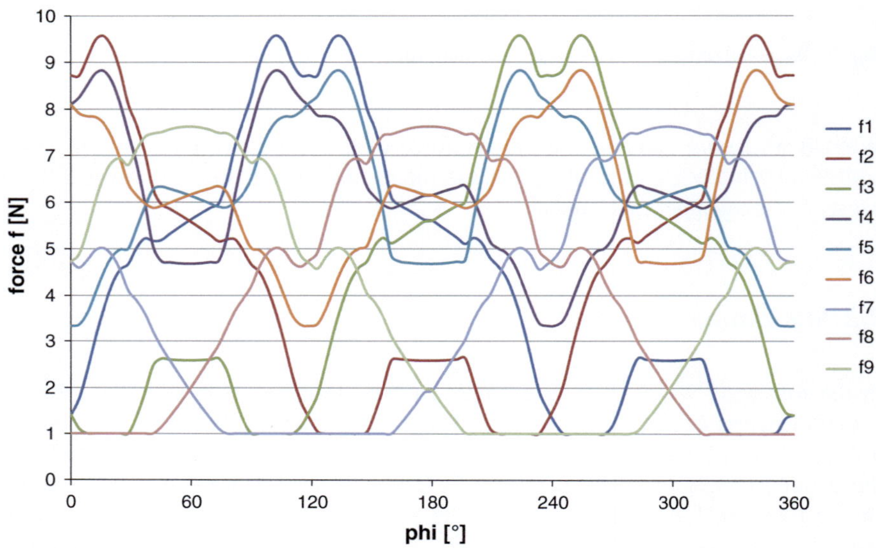

Fig. 4 Run of the nine forces f_i along the z rotation with angle ϕ

4 Conclusion

In this paper new designs of cable-driven parallel robots with cable-driven axes and multiple platforms where introduced. The derived kinematics and structure equations can be used for system design, workspace analysis, and control algorithms. We showed that multi-platform cable robots can be modeled by the same class of equations and structures as cable robots with only one platform which allows the reuse of most algorithms for system analysis and design immediately. In the future the models will be used to address a wider range of applications for cable robots. The demonstrator at Fraunhofer IPA will be used to evaluate different scenarios related to the here proposed robot architectures.

Acknowledgments The research leading to these results has received funding from the European Communitys Seventh Framework Programme under grant agreement No. NMP2-SL-2011-285404 (CABLEBOT).

References

Gouttefarde M, Merlet J-P, Daney D (2007) Wrench-feasible workspace of parallel cable-driven mechanisms. In: IEEE international conference on robotics and automation. Roma, Italy, pp 1492–1497

Kossowski C, Notash L (2002) CAT4 (Cable actuated truss—4 degrees of freedom): a novel 4 DOF cable actuated parallelmManipulator. J Rob Syst 19:605–615

Manfred H, Shiqing F, Sonja M, Richard V, Daniel F (April 2005) Design, analysis and realization of tendon-based parallel manipulators. Mech Mach Theor 40(4):429–445

Merlet J-P, Daney D (2007) A new design for wire-driven parallel robot. In: 2nd international congress, design and modelling of mechanical systems

Ming A, Higuchi T (1994) Study on multiple degree-of-freedom positioning mechanism using wires (Part 1)—concept, design and control. Int J Jap Soc Precis Eng 28(2):131–138

Pott A, Bruckmann T, Mikelsons L (2009) Closed-form force distribution for parallel wire robots. In: Computational kinematics. Springer, Duisburg, Germany, pp 25–34

Varziri MS, Notash L (2007) Kinematic calibration of a wire-actuated parallel robot. Mech Mach Theor 8:960–976

Verhoeven R (2004) Analysis of the Workspace of Tendon-based Stewart Platforms. PhD thesis, University of Duisburg-Essen, Duisburg

Two Gradient-Based Control Laws on $SE(3)$ Derived from Distance Functions

Jan Maximilian Montenbruck, Gerd S. Schmidt, Andrés Kecskeméthy and Frank Allgöwer

Abstract We are interested in gradient systems on the special Euclidean group with application in the control of rigid bodies. This is motivated by the idea of lifting the gradient system to a control law for a systems with Newtonian dynamics, all in the spirit of Daniel Koditschek. In particular, we want to compute gradients of distance functions; in these flows, we can enforce stability of our reference configurations by construction. Therefore, we first outline the computation of a gradient systems on $SE(3)$ on the example of a distance function associated with a Riemannian metric proposed by Frank Park and Roger Brockett. Consequently, we choose a distance function that is easy to compute in camera vision systems and derive the corresponding gradient flow.

We are indebted to the German Research Foundation (DFG) for financial support of the project within the Cluster of Excellence in Simulation Technology (EXC 310/1) at the University of Stuttgart and to Daniel Peralta-Salas for fruitful discussions.

J.M. Montenbruck (✉) · G.S. Schmidt · F. Allgöwer
Institute for Systems Theory and Automatic Control, University of Stuttgart, Stuttgart, Germany
e-mail: jan-maximilian.montenbruck@ist.uni-stuttgart.de

G.S. Schmidt
e-mail: gerd.schmidt@ist.uni-stuttgart.de

F. Allgöwer
e-mail: frank.allgower@ist.uni-stuttgart.de

A. Kecskeméthy · J.M. Montenbruck
Chair of Mechanics and Robotics, University of Duisburg-Essen, Duisburg, Germany
e-mail: andres.kecskemethy@uni-due.de

A. Kecskeméthy and F. Geu Flores (eds.), *Interdisciplinary Applications of Kinematics*, Mechanisms and Machine Science 26, DOI 10.1007/978-3-319-10723-3_4

1 Introduction

The Lie group $SE(3)$ is of special interest in various applications; amongst them are camera positioning, vehicle trajectory planning, and robot modeling, to name a few. For some of these problems, it is important to generate curves on $SE(3)$ (planning problem). For others, one wants to move a point on $SE(3)$ to another (control problem). Both can essentially be described as the problem of minimizing distances on $SE(3)$ (offline for the former and online for the latter case).

Here, we want to consider special gradient algorithms based on distance functions to solve planning and control problems on $SE(3)$. More precisely, we consider state feedback control laws for the kinematic equations of a rigid body. The consideration of the kinematic equations is not restrictive in the considered setup, since there are methods in literature to derive a controller for the full equations of motion from the control law for the kinematic equations (Koditschek and Rimon 1990). We derive the state feedback laws with the help of distance functions. The resulting closed loop vector field is then given by the gradient of the respective distance function, which permits the analysis of the closed loop convergence behavior. We utilize two distance functions. The first distance function is associated with the Riemannian metric proposed by Park and Brockett (1994). the second distance function is one that is particularly easy to compute in camera vision systems.

When choosing coordinate charts, there are established solutions to the above problems (Murray et al. 1994). That is, one chooses an appropriate local parametrization, for instance Euler angles or quaternions, to then implement known control procedures in these coordinates. Only, given this, one has to implement a rule that applies when switching coordinate charts. In contrast, working without coordinate charts can be of interest whether because the resulting methods can be more *objective* (Lin and Burdick 2000) or just more *natural* (Wen and Kreutz-Delgado 1991; Koditschek 1988). In the past, gradient systems have been used when working without coordinate charts; in particular, gradients of Morse-Bott functions have provided feedback laws with almost global convergence (Cunha et al. 2008; Schmidt et al. 2012, 2013).

The remainder of the paper is structured as follows; Sect. 2 entails some preliminaries and the problem formulation, where we explain some basic facts about the special Euclidean group in Sect. 2.1 and state the problem in Sect. 2.2. In Sect. 3, we present our main results. Therein, Sect. 3.1 contains the design procedure for the control law based on the scale-dependent metric and Sect. 3.2 contains the design procedure for the control law based on distance function from camera vision. Within both subsections, we split the design procedure into three subsubsections; Sects. 3.1.1 and 3.2.1 contain the formulations for the distance functions, respectively. In Sects. 3.1.2 and 3.2.2, we derive the corresponding gradients. Consequently, we propose the associated control laws in Sects. 3.1.3 and 3.2.3 and investigate their convergence properties. We conclude the paper in Sect. 4.

2 Preliminaries and Problem Statement

In this section, we briefly sketch some facts about the special Euclidean group as well as the problem under investigation.

2.1 Preliminaries

In the following, we collect some background information on the special Euclidean group $SE(3)$. For a detailed exposition of the following we refer to Murray et al. (1994, Chap. 2). The special Euclidean group is the set $SE(3) = \{(R, d) | R \in SO(3), d \in \mathbb{R}^3\}$ together with the group operation $(R_1, d_1)(R_2, d_2) \mapsto (R_1 R_2, R_1 d_2 + d_1)$, where $SO(3) = \{R \in \mathbb{R}^{3\times3} | R^{-1} = R^\top, \det R = 1\}$ is the set of rotation matrices. The tangent space of $SO(3)$ at a point $R \in SO(3)$ is given by $\mathrm{T}_R\, SO(3) = \{\xi \in \mathbb{R}^{3\times3} | \xi = R\Omega, R \in SO(3), \Omega = -\Omega^\top\}$. The Lie algebra $\mathfrak{so}(3)$ of $SO(3)$ is given by $\mathfrak{so}(3) = \mathrm{T}_I\, SO(3)$, which are the skew-symmetric matrices. For a $\mathbb{R}^3$ association of a $\mathfrak{so}(3)$ element, we can use the natural function $Q : \mathfrak{so}(3) \to \mathbb{R}^3$ given through $Q(\Omega) \times x = \Omega x$, where $\times$ is the cross-product. As a consequence, the Lie algebra $\mathfrak{se}(3)$ of $SE(3)$ is given by (Ω, v) where $\Omega \in \mathfrak{so}(3)$ and $v \in \mathbb{R}^3$.

A compact and common notation for the elements of $SE(3)$ which we also utilize here is the so-called homogeneous representation, where we write tuples (R, d) as matrices H given through

$$H = \begin{bmatrix} R & d \\ 0 & 1 \end{bmatrix} \in SE(3)\,. \tag{1}$$

The group operation then corresponds to matrix multiplication. In a similar fashion, we can also represent the element $(\Omega, v) \in \mathfrak{se}(3)$ as matrices, i.e.

$$T = \begin{bmatrix} \Omega & v \\ 0 & 0 \end{bmatrix} \in \mathfrak{se}(3)\,. \tag{2}$$

By this, the tangent space of $SE(3)$ at a point $H \in SE(3)$ is given by $\mathrm{T}_H\, SE(3) = \{V \in \mathbb{R}^{4\times4} | V = HT, H \in SE(3), T \in \mathfrak{se}(3)\}$, when using matrix notation, and we will refer to its elements as

$$V = \begin{bmatrix} \xi & \zeta \\ 0 & 0 \end{bmatrix} \in \mathrm{T}_H\, SE(3)\,. \tag{3}$$

Thus, $SE(3)$ is invariant with respect to every dynamical system of form $\dot{H} = HT$ with $H \in SE(3)$ and $T \in \mathfrak{se}(3)$. One refers to such equations as the kinematic equations and T underlies a dynamical system itself, which is referred to as the dynamic equations. Most generally, the control input is applied to these dynamic equations. However, one can as well assume to have T as the control input as one can derive a suitable input for the dynamic equations for every choice of T (Koditschek and Rimon 1990).

2.2 Problem Statement

We consider control systems on $SE(3)$ of the form

$$\dot{H} = HU, \tag{4}$$

where $H \in SE(3)$ is the state of the system and $U \in \mathfrak{se}(3)$ is an element of the Lie algebra, making $\dot{H}$ an element of $\mathrm{T}_H SE(3)$. The problem we want to solve is the following; find a state-feedback law of the form

$$U = U\left(H, H^*\right), \tag{5}$$

such that the closed loop converges to the reference $H^* \in SE(3)$ for almost any initial condition $H \in SE(3)$ and such that H^* is stable. Although mechanical control systems usually entail Newtonian dynamics, the above control problem is of interest for such systems (Koditschek and Rimon 1990). This is because one can derive a control law for the system with Newtonian dynamics from the control law for the system with dynamics (4).

3 Main Results

In the following we want to derive state feedback laws to solve the problem described in Sect. 2.2. We utilize a three-step procedure to derive the feedback law. In the first step, we define a distance function which measures the distance between our initial configuration H and the desired configuration H^*. In the second step, we compute the differential of this distance function which we utilize in the third step to derive a feedback (5) such that the closed loop vector field is the gradient of the distance function. In the subsequent discussion, we show that the resulting closed loop has the desired convergence properties. We carry out these computations for two different distance functions which result in two different closed loop systems. The first one is discussed in Sect. 3.1 and the second one in Sect. 3.2.

3.1 A Gradient-Based Controller Based on the Scale-Dependent Metric

This subsection is dedicated to a control law derived from the gradient of a particular distance function based on the left-invariant Riemannian metric proposed by

Frank Park and Roger Brockett, commonly referred to as the scale-dependent metric. We subsequently derive the distance function, the corresponding gradient, and the associated control law.

3.1.1 The Distance Function for the Scale-Dependent Metric

In this section, we compute the distance function based on the Riemannian metric proposed in Park and Brockett (1994). The metric structure both of $SO\,(3)$ and $SE\,(3)$ is of interesting nature and has been investigated excessively, especially by Park and Brockett (1994); Park (1995), among others. In particular, left-invariant Riemannian metrics are of interest in application, as they are independent of inertial coordinates (Lin and Burdick 2000), which are the coordinates placing $\mathbb{R}^3$ in $\mathbb{E}^3$. Herein, we will thus rely on the left-invariant Riemannian metric $\langle \cdot, \cdot \rangle : \mathrm{T}_H\, SE\,(3) \times \mathrm{T}_H\, SE\,(3) \to \mathbb{R}$ proposed by Park and Brockett (1994) given through

$$\langle V, V^* \rangle = \begin{bmatrix} Q\left(R^\top \xi\right) \\ R^\top \zeta \end{bmatrix}^\top \begin{bmatrix} \alpha I & 0 \\ 0 & \beta I \end{bmatrix} \begin{bmatrix} Q\left(R^\top \xi^*\right) \\ R^\top \zeta^* \end{bmatrix}, \tag{6}$$

where $\begin{bmatrix} Q\left(R^\top \xi\right) & R^\top \zeta \\ 0 & 0 \end{bmatrix} = T$ and $\begin{bmatrix} Q\left(R^\top \xi^*\right) & R^\top \zeta^* \\ 0 & 0 \end{bmatrix} = T^*$; The procedure of bringing tangent elements to the identity by group multiplication and then calculating the Riemannian metric with elements of the Lie algebra is common on Lie groups. In mechanics, the resulting notion is called the *twist* $T = H^{-1} V$ ($H \in SE\,(3)$, $V \in \mathrm{T}_H\, SE\,(3)$, and $T \in \mathfrak{se}\,(3)$). The Riemannian metric (6) is called the *scale dependent metric*. This if of interest as the scale-dependence vanishes as we progress with our design procedure, an effect that can be interesting in practice (Angeles 2006; Angeles et al. 1992; Ranjbaran et al. 1996).

The geodesics $\Gamma : \mathbb{R} \to SE\,(3)$, $s \mapsto \Gamma\,(s)$ between H and H^* associated with (6) are found by minimizing the functional $\int_0^1 \langle \frac{\mathrm{d}}{\mathrm{d}s}\Gamma, \frac{\mathrm{d}}{\mathrm{d}s}\Gamma \rangle\, \mathrm{d}s$ over all curves joining H and H^* such that $\Gamma\,(0) = H$ and $\Gamma\,(1) = H^*$. Calculating geodesics can in general turn out to be tedious and is not within the scope of this paper. We therefore omit the precise calculation and instead refer to Park (1995) for details. To sketch the calculation, we only want to mention that geodesics on $SE\,(3)$ between H and H^* associated with the Riemannian metric (6) can be obtained from the geodesics in $SO\,(3)$ and $\mathbb{R}^3$, yielding

$$\Gamma\,(s) = \begin{bmatrix} R \exp\left(\log\left(R^\top R^*\right) s\right) & d + s\,(d^* - d) \\ 0 & 1 \end{bmatrix}, \tag{7}$$

where $\log : SO\,(3) \to \mathfrak{so}\,(3)$ and $\exp : \mathfrak{so}\,(3) \to SO\,(3)$ are the logarithmic and the exponential map, respectively. Notably, (7) is the one-parameter family of screw motions.

The twist T of Γ is given through the formula

$$T(s) = \Gamma(s)^{-1} \frac{\mathrm{d}}{\mathrm{d}s} \Gamma(s), \tag{8}$$

and we have

$$T(s) = \left(\log\left(R^\top R^*\right), \left(\exp\left(\log\left(R^\top R^*\right) s\right)\right)^\top R^\top \left(d^* - d\right) \right). \tag{9}$$

Applying the Riemannian metric (6) to the twist (9), we have

$$\langle \frac{\mathrm{d}\Gamma(s)}{\mathrm{d}s}, \frac{\mathrm{d}\Gamma(s)}{\mathrm{d}s} \rangle = \alpha \left(Q\left(\log\left(R^\top R^*\right)\right)\right)^\top Q\left(\log\left(R^\top R^*\right)\right) + \beta \left(d^* - d\right)^\top \left(d^* - d\right),$$

which we integrate over the interval $[0, 1]$. Then, applying the useful identity $2Q(\Omega)^\top Q(\Omega) = \operatorname{tr}\left(\Omega^\top \Omega\right)$ for $\Omega \in \mathfrak{so}(3)$, we arrive at our distance function $d : SE(3) \times SE(3) \to \mathbb{R}$ given by

$$d^2\left(H, H^*\right) = \frac{\alpha}{2} \operatorname{tr}\left(\left(\log\left(R^\top R^*\right)\right)^\top \log\left(R^\top R^*\right)\right) + \beta \left(d^* - d\right)^\top \left(d^* - d\right), \tag{10}$$

which just agrees fine with the distance function obtained in Park (1995).

3.1.2 The Gradient for the Scale-Dependent Metric

We are hence in the position to describe our error function $e : SE(3) \to \mathbb{R}$ as the distance between H and H^*, meaning e is the same function as d, only for fixed H^*. This is just writing $e(H) = d^2(H, H^*)$. We can therefore formulate our design goal as the the optimization problem

$$\begin{aligned} &\text{minimize} && e(H) \\ &\text{subject to} && H \in SE(3). \end{aligned} \tag{11}$$

Next, as we have taken e as our "cost", we have to compute the direction in which e decreases, i.e. the tangent element of $SE(3)$ which is gradient of e. For doing so, we first need the directional derivative of e at $V \in \mathrm{T}SE(3)$, which is

$$\mathrm{d}_H e(V) = \left. \frac{\mathrm{d}}{\mathrm{d}\gamma} e \circ A(\gamma) \right|_{\gamma=0}, \tag{12}$$

where $A : [-\varepsilon, \varepsilon] \to SE(3)$ is such that $A(0) = H$ and $\frac{\mathrm{d}}{\mathrm{d}\gamma} A(\gamma) = V$. Writing $A(\gamma) = (R_A(\gamma), d_A(\gamma))$ and $V = (\xi, \zeta)$, we have

$$\alpha \operatorname{tr}\left(\left(\log\left(R_A^\top(\gamma)\, R^*\right)\right)^\top R^{*\top} R_A(\gamma) \frac{\mathrm{d}}{\mathrm{d}\gamma} R_A^\top(\gamma)\, R^*\right) + 2\beta \left(d^* - d_A(\gamma)\right)^\top \left(-\frac{\mathrm{d}}{\mathrm{d}\gamma} d_A(\gamma)\right)\Big|_{\gamma=0}$$

$$= \alpha \operatorname{tr}\left(\left(\log\left(R^\top R^*\right)\right)^\top R^{*\top} R \xi^\top R^*\right) + 2\beta \left(d^* - d\right)^\top (-\zeta) = \mathrm{d}_H\, e(V). \tag{13}$$

To obtain the gradient, one has to apply the formula

$$\langle \operatorname{grad} e(H), V \rangle = \mathrm{d}_H\, e(V), \tag{14}$$

where V is assumed to be tangent to the same element, that $\operatorname{grad} e(H)$ is tangent to, and to solve for $\operatorname{grad} e(H)$. Using the representation $\operatorname{grad} e(H) = \left(\xi_{\text{grad}}, \zeta_{\text{grad}}\right)$, we arrive at

$$\frac{\alpha}{2} \operatorname{tr}\left(\xi_{\text{grad}}^\top \xi\right) + \beta \zeta_{\text{grad}}^\top \zeta = \mathrm{d}_H\, e(V). \tag{15}$$

Knowing that the trace is invariant under both, cyclic permutations and transposing, and applying the rule $(\log(R))^\top = -\log(R)$ for $R \in SO(3)$, we are able to equate coefficients between (13) and (15) to get

$$\operatorname{grad} e(H) = \begin{bmatrix} -2R^* \log\left(R^\top R^*\right) R^{*\top} R & -2\left(d^* - d\right) \\ 0 & 0 \end{bmatrix}. \tag{16}$$

3.1.3 The Control Law for the Scale-Dependent Metric

We arrive at the dynamical system $\dot{H} = -\operatorname{grad} e(H)$. That is

$$\dot{H} = \begin{bmatrix} 2R^* \log\left(R^\top R^*\right) R^{*\top} R & 2\left(d^* - d\right) \\ 0 & 0 \end{bmatrix}, \tag{17}$$

similarly to the results of Bullo and Murray (1995), and we will consequently investigate the stability of (17).

Remark 1 If we want to include time-dependence of H^* explicitly (in the sense that it is a reference signal), and if $H^*(t)$ is sufficiently smooth, we could repeat the last steps of our derivation to get

$$\dot{H} = \begin{bmatrix} 2R^* \log\left(R^\top R^*\right) R^{*\top} R - R^* \dot{R}^{*\top} R & 2\left(d^* - d\right) + \dot{d}^* \\ 0 & 0 \end{bmatrix} \tag{18}$$

instead of (17).

Theorem 1 *The equilibrium $H = H^*$ of* (17) *is asymptotically stable.*

Proof Suppose the Lyapunov function candidate $W(H) = e(H)$. We have W positive semidefinite because d is a distance function. Further, W is zero iff $H = H^*$. We take the directional derivative

$$\dot{W}(H) = \frac{\alpha}{2}\operatorname{tr}\left(\left(R^{*\top}R\dot{R}^{\top}R^{*}\right)^{\top}\log\left(R^{\top}R^{*}\right) + \left(\log\left(R^{\top}R^{*}\right)\right)^{\top}R^{*\top}R\dot{R}^{\top}R^{*}\right) \\ - 2\beta\dot{d}^{\top}\left(d^{*}-d\right),$$

and, substituting (17), we have

$$\dot{W}(H) = 2\alpha\operatorname{tr}\left(\left(\log\left(R^{\top}R^{*}\right)\right)^{2}\right) - 4\beta\left(d^{*}-d\right)^{\top}\left(d^{*}-d\right), \tag{19}$$

which equals $\dot{W}(H) = -4W(H)$ and means that $\dot{W}(H)$ is negative semidefinite and zero if $H = H^{*}$. □

3.2 A Gradient-Based Controller Based on a Distance Function from Camera Vision

Again, we split the subsection into three parts. First, we define our distance function. Then, we take the gradient with respect to one of its arguments. Consequently, we define our control law accordingly and investigate its convergence properties.

3.2.1 The Distance Function from Camera Vision

We could see that the gradient flow of the distance function (10) computed above had some nice convergence properties. However, to compute the distance function (10) one has to compute the rotation matrices of the current and the desired configuration, respectively. Instead, in camera vision systems, it is common to only know the *position* of some characteristic points of the rigid body. These points are usually captured with camera markers, e.g. retroflective or colored markers. In such settings, the distance between the current and the desired position of the markers

$$d^{2}\left(H, H^{*}\right) = \sum_{b\in M} b^{\top}\left(R-R^{*}\right)^{\top}\left(R-R^{*}\right)b + \left(d-d^{*}\right)^{\top}\left(d-d^{*}\right) \tag{20}$$

where M is the set of marker positions in body-fixed coordinates, appears to be an appropriate distance function. In particular, it approximates the volume enclosed by the body particles between current and desired position

$$d^{2}\left(H, H^{*}\right) = \int_{B} b^{\top}\left(R-R^{*}\right)^{\top}\left(R-R^{*}\right)b\,\mathrm{d}b + \left(d-d^{*}\right)^{\top}\left(d-d^{*}\right), \tag{21}$$

where B is the set of body particles in body-fixed coordinates. We now mimic the steps taken to arrive at (17).

3.2.2 The Gradient for the Distance Function from Camera Vision

First, we define $e(H) = d^2(H, H^*)$ and formulate the optimization problem (11) to then compute the directional derivative

$$\mathrm{d}_H e(V) = -\int_B b^\top \left(R^{*\top}\xi + \xi^\top R^*\right) b \,\mathrm{d}b + 2\left(d - d^*\right)^\top \zeta \qquad (22)$$

and apply $\mathrm{d}_H e(V) = \langle \operatorname{grad} e(H), V \rangle$. We thus have

$$-\int_B b^\top \left(R^{*\top}\xi + \xi^\top R^*\right) b \,\mathrm{d}b + 2\left(d - d^*\right)^\top \zeta = \alpha Q\left(R^\top \xi_{\mathrm{grad}}\right)^\top Q\left(R^\top \xi\right) + \beta\left(R^\top \zeta_{\mathrm{grad}}\right)^\top \left(R^\top \zeta\right). \qquad (23)$$

Equating coefficients for ζ, we have $\zeta_{\mathrm{grad}} = \frac{2}{\beta}(d - d^*)$. Equating what remains, we arrive at

$$-\int_B b^\top \left(R^{*\top}\xi + \xi^\top R^*\right) b \,\mathrm{d}b = \frac{\alpha}{4} \operatorname{tr}\left(\xi_{\mathrm{grad}}^\top \xi + \xi^\top \xi_{\mathrm{grad}}\right), \qquad (24)$$

where we have used the identities $2Q(\Omega_1)^\top Q(\Omega_2) = \operatorname{tr}\left(\Omega_1^\top \Omega_2\right)$, $\Omega_1, \Omega_2 \in \mathfrak{so}(3)$ and $2\operatorname{tr}(A) = \operatorname{tr}\left(A + A^\top\right)$. We now suppose that ξ and ξ_{grad} are both tangent to R and hence use the ansatz $\xi = R\Omega$, $\xi_{\mathrm{grad}} = R\Omega_{\mathrm{grad}}$. This yields

$$\int_B b^\top \left(\Omega R^\top R^* - R^{*\top} R \Omega\right) b \,\mathrm{d}b = -\frac{\alpha}{4} \operatorname{tr}\left(\Omega_{\mathrm{grad}}\Omega + \Omega\Omega_{\mathrm{grad}}\right). \qquad (25)$$

It follows by some tedious computations that

$$\Omega_{\mathrm{grad}} = \frac{2}{\alpha} \int_B \Omega_1 b^\top \Omega_1 R^\top R^* b + \Omega_2 b^\top \Omega_2 R^\top R^* b + \Omega_3 b^\top \Omega_3 R^\top R^* b \mathrm{d}b \qquad (26)$$

satisfies (25) for all $\Omega \in \mathfrak{so}(3)$, where $\Omega_1, \Omega_2, \Omega_3$ are the generators of the algebra $\mathfrak{so}(3)$ given by $\Omega_1 = \begin{bmatrix} 0 & 0 & 0 \\ 0 & 0 & -1 \\ 0 & 1 & 0 \end{bmatrix}$, $\Omega_2 = \begin{bmatrix} 0 & 0 & 1 \\ 0 & 0 & 0 \\ -1 & 0 & 0 \end{bmatrix}$, and $\Omega_3 = \begin{bmatrix} 0 & -1 & 0 \\ 1 & 0 & 0 \\ 0 & 0 & 0 \end{bmatrix}$. Consistency between (25) and (26) can however be checked by substituting (26) back into (25). This turns out to be true for all $\Omega \in \mathfrak{so}(3)$.

3.2.3 The Control Law for the Distance Function from Camera Vision

Interestingly, the gradient flow of (21) is, in contrast to the gradient flow of (10), *not* scale-independent. Instead of (17), we hence have

$$\dot{H} = \begin{bmatrix} -\frac{2}{\alpha} R \int_B \sum_{i=1}^{3} \Omega_i b^\top \Omega_i R^\top R^* b \,\mathrm{d}b & -\frac{2}{\beta}\left(d - d^*\right) \\ 0 & 0 \end{bmatrix} \tag{27}$$

and we are consequently interested in the stability of the equilibrium $H = H^*$ of (27).

Theorem 2 *The equilibrium $H = H^*$ of* (27) *is asymptotically stable.*

Proof Consider the Lyapunov function candidate $W(H) = e(H)$. W is positive semidefinite because e is a distance function and $W(H) = 0$ iff $H = H^*$. Now, taking the directional derivative, we have

$$\dot{W}(H) = \int_B -b^\top \left(R^{*\top} \dot{R} + \dot{R}^\top R^*\right) b \,\mathrm{d}b + 2\left(d - d^*\right) \dot{d}. \tag{28}$$

Substituting (27), this is

$$\dot{W}(H) = \frac{2}{\alpha} \int_B b^\top \left(R^{*\top} R\left(\sum_{i=1}^{3} \Omega_i b^\top \Omega_i R^\top R^* b\right) + \left(\sum_{i=1}^{3} \Omega_i b^\top \Omega_i R^\top R^* b\right)^\top R^\top R^*\right) b \,\mathrm{d}b - \frac{4}{\beta}\left(d - d^*\right)^\top \left(d - d^*\right)$$

and we substitute $b^\top R^{*\top} R \Omega_i b = \xi_i$ to see that the above is in fact

$$\dot{W}(H) = -\frac{2}{\alpha} \int_B 2\xi_1^2 + 2\xi_2^2 + 2\xi_3^2 \,\mathrm{d}b - \frac{4}{\beta}\left(d - d^*\right)\left(d - d^*\right) \tag{29}$$

which satisfies $\dot{W}(H) \leq 0$ and $\dot{W}(H) = 0$ if $H = H^*$. □

4 Conclusion

Inspired by the special properties of distance functions we have computed two gradient systems on the special Euclidean group with the intention to use them as control laws for rigid bodies under Newtonian dynamics. In the first case, we have chosen a distance function that we derived from the scale-dependent metric of Frank Park and Roger Brockett. We found that the resulting system has an asymptotically stable equilibrium at the reference configuration. Subsequently, we mimicked this approach with a distance function that is particularly suited for computation in camera vision systems. Again, we could find that the corresponding gradient system had nice convergence properties; the reference configuration is an asymptotically stable equilibrium.

Open topics include the reduction of the number of feedback variables, inclusion of joint and workspace constraints, as well as the formulation of our control laws for systems under Newtonian dynamics.

References

Murray RM, Li Z, Sastry SS (1994) A mathematical introduction to robotic manipulation. CRC Press, Berkeley

Lin Q, Burdick J (2000) On well-defined kinematic metric functions. In: Proceedings of the 2000 Internatoinal Conference on Robotics and Automation, pp 170–177

Wen JT-Y, Kreutz-Delgado K (1991) The attitude control problem. IEEE Trans Autom Control 36:1148–1162

Koditschek DE (1988) Application of a new Lyapunov function to global attitude tracking. In: Proceedings of 27th conference decision and control, pp 63–68

Koditschek DE, Rimon E (1990) Robot navigation functions on manifolds with boundary. Adv Appl Math 11:412–442

Park FC, Brockett RW (1994) Kinematic dexterity of robotic mechanisms. Int J Robot Res 13(1):1–15

Cunha R, Silvestre C, Hepanha J (2008) Output-feedback control for stabilization on SE(3). Syst Control Lett 57:1013–1022

Schmidt GS, Ebenbauer C, Allgöwer F (2012) A solution for a class of output regulation problems on SO(n). In: Proceedings of American control conference 2012, pp 1773–1779

Schmidt GS, Ebenbauer C, Allgöwer F (2013) Output regulation for attitude control: a global approach. In: Proceedings of 2013 American control conference

Schmidt GS, Ebenbauer C, Allgöwer F (2009) Global output regulation for the rotational dynamics of a rigid body. Automatisierungstechnik

Park FC (1995) Distance metrics on the rigid-body motions with applications to mechanism design. ASME J Mech Des 117:48–54

Angeles J (2006) Is there a characteristic length of a rigid-body displacement? Mech Mach Theory 41:884–896

Angeles J, Ranjbaran F, Patel R (1992) On the design of the kinematic structure of seven-axes redundant manipulators for maximum conditioning. In: Proceedings of 1992 international conference robotics and automation, pp 494–499

Ranjbaran F, Angeles J, Kecskemethy A (1996) On the kinematic conditioning of robotic manipulators. In: Proceedings of 1996 international conference robotics and automation, pp 3167–3172

Bullo F, Murray RM (1995) Proportional derivative (pd) control on the Euclidean group. California Institute of Technology, Technical report

Invariant Properties of the Denavit–Hartenberg Parameters

Mohammed Daher and Peter Donelan

Abstract The Denavit–Hartenberg (DH) notation for kinematic chains makes use of a set of parameters that determine the relative positions of and between successive joints. The corresponding matrix representation of a chain's kinematics is a product of two exponentials in the homogeneous representation of the Euclidean group. While the DH notation is based on sound kinematic intuition, it is not obviously natural in mathematical terms. In this paper, we use the principle of transference to determine fundamental algebraic (polynomial) invariants of the Euclidean group $SE(3)$ acting on sets of twists, elements of the group's Lie algebra, $\mathfrak{se}(3)$, representing joints, and show that the DH parameters are algebraic functions of these invariants. We make use of the fact that for a set of three twists, there is an algebraic–geometric duality with the corresponding set of Lie brackets, so that link lengths of one correspond to offsets of the other.

Keywords Euclidean group · Denavit–Hartenberg parameters · Polynomial invariants · Dual numbers

1 Introduction

The kinematics of serial chains are usually modelled either by the Denavit–Hartenberg notation (Denavit and Hartenberg 1955) or by the product of exponentials notation. In fact, Brockett (1984) showed how the two are related to one another. The key idea in both cases is that the motion of a link with respect to its neighbour in the chain is described by an exponential matrix in terms of the joint parameter. This arises from the fact that the one degree-of-freedom joints—revolute, helical or prismatic—can be represented by a twist (or motor or screw). In mathematical terms, a twist

M. Daher · P. Donelan (✉)
School of Mathematics, Statistics and Operations Research, Victoria University of Wellington, Wellington, New Zealand
e-mail: peter.donelan@vuw.ac.nz

M. Daher
e-mail: Mohammed.Daher@msor.vuw.ac.nz

A. Kecskeméthy and F. Geu Flores (eds.), *Interdisciplinary Applications of Kinematics*, Mechanisms and Machine Science 26, DOI 10.1007/978-3-319-10723-3_5

is an element of the Lie algebra $\mathfrak{se}(3)$ of the Euclidean group $SE(3)$ of rigid-body transformations.

Twists can be represented in a variety of ways, but the most succinct is using Plücker coordinates (Selig 2005). We shall write a twist S as a pair of 3-vectors: $S = (\boldsymbol{\omega}, \mathbf{v})$. These coordinates rely on a choice of spatial coordinate frame, so we would expect kinematic properties to be invariant under changes of coordinate. Mathematically, they should be invariants of the *adjoint action* Ad of the Euclidean group on its Lie algebra. The most fundamental of these is the *pitch* of a twist, which in terms of Plücker coordinates is the ratio:

$$h = \frac{\boldsymbol{\omega}.\mathbf{v}}{\boldsymbol{\omega}.\boldsymbol{\omega}} \tag{1}$$

of the Klein form $\boldsymbol{\omega}.\mathbf{v} = \omega_1 v_1 + \omega_2 v_2 + \omega_3 v_3$ and Killing form $\boldsymbol{\omega}.\boldsymbol{\omega} = \omega_1^2 + \omega_2^2 + \omega_3^2$. Each of these is in fact an *invariant* polynomial $f(\boldsymbol{\omega}, \mathbf{v})$ in the sense that for any $g \in SE(3)$:

$$f(Ad(g)(\boldsymbol{\omega}, \mathbf{v})) = f(\boldsymbol{\omega}, \mathbf{v}). \tag{2}$$

For a serial chain, we have more than one joint, and hence are interested in the invariants of a set of twists $S_1, \ldots, S_k$. Since the Euclidean group is algebraic, that is it can be represented as the zero set of polynomials, its polynomial invariants are of particular significance. Other invariant quantities, such as the pitch, may be expressed as rational or algebraic in terms of them. Our primary goal is to determine fundamental polynomial invariants for sets of twists and to show that the DH parameters can indeed be written in terms of them.

Our guide is the *principle of transference*, whose origins are in the work of Clifford (Rooney 2007) and whose value was recognised by Kotelnikov (1895) and, especially, Study (1903)—see, for example, Chevallier (1996), Rico Martínez and Duffy (1993), Rooney (1975), Selig (2005). An algebraic version of the principle states that on replacing real coordinates by dual coordinates, valid statements about vectors in $\mathbb{R}^3$ become valid statements about twists, written as dual vectors $\boldsymbol{\omega} + \varepsilon\mathbf{v}$, where ε is a quantity such that $\varepsilon^2 = 0$. Chevallier notes that this should not be regarded as a theorem, as there are exceptions to its application: it is a valuable generic guide.

We follow Study (1903), Sect. 23, in making use of the principle by starting with invariants and syzygies of the rotation group $SO(3)$, acting on m copies of $\mathbb{R}^3$ (Weyl 1997), which we refer to as m-fold invariants. Dualisation in $SO(3)$ leads to dual invariant polynomials, whose real and dual parts are real invariants of the Euclidean group. A number of authors (Rosen 1968; Perroud 1983; Takiff 1972; Donelan and Gibson 1991) have explored invariants of the adjoint and co-adjoint action (the latter of importance in theoretical physics). Selig (2005) also makes use of the principle of transference to derive invariants that correspond to those we obtain. In Crook (2009), the algebraic method of SAGBI bases is employed to find some of these invariants. The work of Takiff (1972) has been extended by Panyushev (2007), but in a theoretical algebraic–geometry setting, and he too obtains the dualised form of 2-fold invariants.

The group and dualisation are described in Sect. 2, while its application to invariants is in Sect. 3. In Sect. 4, the Denavit–Hartenberg parameters are derived in terms of Plücker coordinates. Since three joints are required to define the offset, the expectation is that it can be written in terms of 3-vector polynomial invariants of the Euclidean group (using Weyl's term). To achieve such as expression we introduce in Sect. 5 the algebraic-geometric duality between a set of three screws and its associated set of Lie brackets.

2 The Euclidean Group and Dualisation

Let $\mathbb{D}$ denote the ring of *dual numbers* $a+\varepsilon b$, $a, b \in \mathbb{R}$ and $\varepsilon^2 = 0$ with componentwise addition, and multiplication defined in the obvious way. Note that $\mathbb{D}$ is not a field, as there are zero divisors and not every non-zero quantity has a multiplicative inverse, but is a 2-dimensional real associative algebra. In a dual number $a + \varepsilon b$, a is referred to as the *primal part* and b, the *dual part*. Modules of various sorts can be constructed by taking vectors and matrices of dual numbers; these can also be written as a sum of primal and dual parts.

The position of a link in a serial chain with respect to some reference position, in a given coordinate frame, is represented by an element of the Euclidean group $SE(3)$. The group is a (semi-direct) product of the orientation-preserving rotations $SO(3)$ and translations $\mathbb{R}^3$, and is a 6–dimensional Lie group (Murray et al. 1994; Selig 2005). The rotation group $SO(3)$ is characterised by the following conditions on a 3×3 (real) matrix A:

$$AA^t = I, \quad \det A = 1. \tag{3}$$

Replacing entries in A by dual numbers gives rise to a dual matrix $\hat{A} = A_0 + \varepsilon A_1$ where A_0, A_1 are real 3×3 matrices, the primal and dual parts respectively. The same equations (3) determine a group $SO(3, \mathbb{D})$ (McCarthy 1986). Equating primal and dual parts we obtain:

$$A_0 A_0^t = I, \quad (A_1 A_0^t)^t = -A_1 A_0^t, \quad \det(A_0) = 1, \tag{4}$$

so that $A_0 \in SO(3)$ and $A_1 A_0^t$ is skew symmetric. Identifying this skew-symmetric matrix with a translation vector in $\mathbb{R}^3$ gives rise to an isomorphism between $SO(3, \mathbb{D})$ and $SE(3)$ that is at the heart of the principle of transference.

Correspondingly, the Lie algebra $\mathfrak{se}(3)$ of the Euclidean group is the sum of $\mathfrak{so}(3)$, the infinitesimal rotations and $\mathfrak{t}(3)$, infinitesimal translations. The algebra $\mathfrak{so}(3)$ can be represented by 3×3 skew-symmetric matrices but these in turn can be identified with elements of $\mathbb{R}^3$ in a standard way (Murray et al. 1994). Dualising works here too, in that twists in $\mathfrak{se}(3)$ can be written as dual vectors $S = \boldsymbol{\omega}+\varepsilon\mathbf{v}$, where the primal and

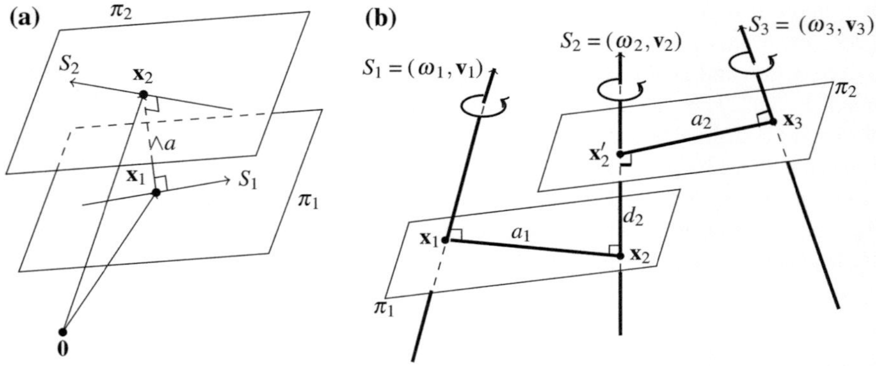

Fig. 1 (a) Link length and (b) offset in Plücker coordinates

dual parts $\boldsymbol{\omega}$, $\mathbf{v}$ are the Plücker coordinates. Geometrically, a twist $S = \boldsymbol{\omega} + \varepsilon\mathbf{v}$ can be interpreted as a vector field on $\mathbb{R}^3$ whose integral curves—the motion generated by the twist—are helices of pitch $h = \boldsymbol{\omega}.\mathbf{v}/\boldsymbol{\omega}.\boldsymbol{\omega}$ about an axis with direction vector $\boldsymbol{\omega}$ and moment $\boldsymbol{\omega} \times \mathbf{q} = \mathbf{v} - h\boldsymbol{\omega}$ about $\mathbf{0}$ (where $\mathbf{q} = (\mathbf{v} \times \boldsymbol{\omega})/\boldsymbol{\omega}.\boldsymbol{\omega}$ is a point on the axis Selig 2005). Note that if $h = 0$ then the the motion is revolute—the integral curves are circles centred on the axis and lying in planes orthogonal to it. If, on the other hand, $\boldsymbol{\omega} = \mathbf{0}$, then the motion is translational and the integral curves are lines parallel to $\mathbf{v}$.

A change of coordinate frame corresponds to conjugation in the group and this gives rise to the Lie bracket, a bilinear, skew-symmetric product, in its Lie algebra (Murray et al. 1994). Writing elements of $\mathfrak{so}(3)$ as 3-vectors $\boldsymbol{\omega}_1, \boldsymbol{\omega}_2 \in \mathbb{R}^3$, the Lie bracket is the standard vector product on $\mathbb{R}^3$, $[\boldsymbol{\omega}_1, \boldsymbol{\omega}_2] = \boldsymbol{\omega}_1 \times \boldsymbol{\omega}_2$ Dualising, so writing elements of $\mathfrak{se}(3)$ as dual vectors $S_i = \boldsymbol{\omega}_i + \varepsilon\mathbf{v}_i$, $i = 1, 2$, the Lie bracket is the dual vector product:

$$[S_1, S_2] = (\boldsymbol{\omega}_1 + \varepsilon\mathbf{v}_1) \times (\boldsymbol{\omega}_2 + \varepsilon\mathbf{v}_2) = \boldsymbol{\omega}_1 \times \boldsymbol{\omega}_2 + \varepsilon(\boldsymbol{\omega}_1 \times \mathbf{v}_2 + \mathbf{v}_1 \times \boldsymbol{\omega}_2). \quad (5)$$

The geometric interpretation of the Lie bracket $[S_1, S_2]$, in the generic case $\boldsymbol{\omega}_1 \times \boldsymbol{\omega}_2 \neq \mathbf{0}$, is as a twist whose axis is the common perpendicular to the axes of S_1, S_2 with twist h_{12} a function of the pitches h_i of S_i and the relative placement of the twists (Samuel et al. 1991).

3 Invariants and Dualisation

The standard action of the rotation group $SO(3)$ on $\mathbb{R}^3$ is the same as its adjoint action on $\mathfrak{so}(3)$. Weyl (1997) gives a complete account of its vector polynomial invariants. They are of two types: given $\boldsymbol{\omega}_1, \ldots, \boldsymbol{\omega}_m \in \mathbb{R}^3 \cong \mathfrak{so}(3)$, every m-fold polynomial invariant is generated by (i.e. is a polynomial function of) the quadratic and cubic invariants

$$I_{ij} = \boldsymbol{\omega}_i.\boldsymbol{\omega}_j, \quad 1 \le i, j \le m; \qquad I_{ijk} = [\boldsymbol{\omega}_i, \boldsymbol{\omega}_j, \boldsymbol{\omega}_k], \quad 1 \le i < j < k \le m \quad (6)$$

The latter is termed a *bracket* and is the determinant of the matrix whose columns are the three vectors. Note that these invariants are not algebraically independent but are linked by several types of *syzygies*, that is, polynomial relations. We shall only be concerned with $m \le 3$ and in these cases the only syzygy occurs when $m = 3$. There are 7 invariants of type (6) connected by a single syzygy

$$I_{123}^2 = \det(I_{ij}). \quad (7)$$

If $f(\boldsymbol{\omega}_1, \ldots, \boldsymbol{\omega}_m)$ is a polynomial in the coordinates of the m vectors, then replacing the real vectors by the dual vectors $\boldsymbol{\omega}_i + \varepsilon \mathbf{v}_i$, $i = 1, \ldots, m$ and expanding by the binomial theorem gives the dual polynomial in differential form:

$$\hat{f}(\boldsymbol{\omega}_1, \ldots, \boldsymbol{\omega}_m, \mathbf{v}_1, \ldots, \mathbf{v}_m) = f(\boldsymbol{\omega}_1, \ldots, \boldsymbol{\omega}_m) + \varepsilon \cdot \sum_{r=1}^{m} \sum_{j=1}^{3} v_{rj} \frac{\partial f}{\partial \omega_{rj}}(\boldsymbol{\omega}_1, \ldots, \boldsymbol{\omega}_m). \quad (8)$$

It is straightforward to show that if f is an m-fold polynomial invariant of $SO(3)$, then the primal and dual parts of $\hat{f}$ are indeed m-fold invariants of the adjoint action of $SE(3)$. For the adjoint action itself, $SO(3)$ has the single generating invariant $I_{11} = \boldsymbol{\omega}.\boldsymbol{\omega}$ which dualises to give:

$$(\boldsymbol{\omega} + \varepsilon \mathbf{v}).(\boldsymbol{\omega} + \varepsilon \mathbf{v}) = \boldsymbol{\omega}.\boldsymbol{\omega} + 2\varepsilon \boldsymbol{\omega}.\mathbf{v}. \quad (9)$$

Up to a multiple, the primal and dual parts I_{11} and $\hat{I}_{11} = \boldsymbol{\omega}.\mathbf{v}$ are the familiar Killing and Klein forms whose ratio is the pitch of the twist $S = (\boldsymbol{\omega}, \mathbf{v})$. For $m = 2$, there are 6 quadratic invariants arising from dualisation of the $SO(3)$ invariants (Crook 2009). In the case $m = 3$, there are 14 invariants arising from the primal and dual parts of the dualisations of the 7 generating 3-fold invariants for $SO(3)$:

$$\begin{aligned}
&I_{ii} = \boldsymbol{\omega}_i \cdot \boldsymbol{\omega}_i, && \hat{I}_{ii} = \boldsymbol{\omega}_i \cdot \mathbf{v}_i, && i = 1, 2, 3 \\
&I_{ij} = \boldsymbol{\omega}_i \cdot \boldsymbol{\omega}_j, && \hat{I}_{ij} = \boldsymbol{\omega}_i \cdot \mathbf{v}_j + \mathbf{v}_i \cdot \boldsymbol{\omega}_j, && 1 \le i < j \le 3 \\
&I_{123} = [\boldsymbol{\omega}_1, \boldsymbol{\omega}_2, \boldsymbol{\omega}_3], && \hat{I}_{123} = [\mathbf{v}_1, \boldsymbol{\omega}_2, \boldsymbol{\omega}_3] + [\boldsymbol{\omega}_1, \mathbf{v}_2, \boldsymbol{\omega}_3] + [\boldsymbol{\omega}_1, \boldsymbol{\omega}_2, \mathbf{v}_3]
\end{aligned} \quad (10)$$

The syzygy (7) can also be dualised and therefore gives rise to a pair of syzygies, the primal part being (7) and the dual part:

$$\begin{aligned}
I_{123}\hat{I}_{123} = &\hat{I}_{11} I_{22} I_{33} - \hat{I}_{11} I_{23}^2 - \hat{I}_{22} I_{13}^2 + \hat{I}_{22} I_{11} I_{33} + \hat{I}_{33} I_{11} I_{22} - \hat{I}_{33} I_{12}^2 - \hat{I}_{12} I_{12} I_{33} \\
&+ \hat{I}_{12} I_{13} I_{23} - \hat{I}_{13} I_{13} I_{22} + \hat{I}_{13} I_{12} I_{23} + \hat{I}_{23} I_{12} I_{13} - \hat{I}_{23} I_{11} I_{23}.
\end{aligned} \quad (11)$$

4 Link Length and Offset

The *link length*—the length of the common perpendicular between the axis-lines of two twists—is one of the DH parameters determined by a successive pair of joints. Let us assume they have finite pitch and the axes of screws are non-parallel (see Fig. (1a)). To calculate the link length d between twists S_1 and S_2 in terms of Plücker coordinates $(\boldsymbol{\omega}_i, \mathbf{v}_i)$, $i = 1, 2$, let plane π_i be the plane that contains the axis for S_i, $i = 1, 2$, and is normal to the common perpendicular; the planes have common normal $\boldsymbol{\omega}_1 \times \boldsymbol{\omega}_2$. The normal vector meets the axes at points:

$$\mathbf{x}_i = \frac{\mathbf{v}_i \times \boldsymbol{\omega}_i}{\boldsymbol{\omega}_i \cdot \boldsymbol{\omega}_i}. \tag{12}$$

The planes have the form $(\boldsymbol{\omega}_1 \times \boldsymbol{\omega}_2) \cdot \mathbf{x} = k_i$, $i = 1, 2$ and the link length is given by $a = (|k_2 - k_1|)\|\boldsymbol{\omega}_1 \times \boldsymbol{\omega}_2\|$. Substituting $\mathbf{x}_i$ from (12) into the plane equations to determine k_i and applying Lagrange's identity:

$$a = \frac{\boldsymbol{\omega}_1 \cdot \boldsymbol{\omega}_2}{\|\boldsymbol{\omega}_1 \times \boldsymbol{\omega}_2\|}\left(\frac{\mathbf{v}_1 \cdot \boldsymbol{\omega}_1}{\boldsymbol{\omega}_1 \cdot \boldsymbol{\omega}_1} + \frac{\mathbf{v}_2 \cdot \boldsymbol{\omega}_2}{\boldsymbol{\omega}_2 \cdot \boldsymbol{\omega}_2}\right) - \frac{\boldsymbol{\omega}_1 \cdot \mathbf{v}_2 + \mathbf{v}_1 \cdot \boldsymbol{\omega}_2}{\|\boldsymbol{\omega}_1 \times \boldsymbol{\omega}_2\|} \tag{13}$$

We observe from (13) that if either $\boldsymbol{\omega}_1$, $\boldsymbol{\omega}_2$, or $\boldsymbol{\omega}_1 \times \boldsymbol{\omega}_2$ is equal to zero, then the link length will be undefined. It is a lengthy but essentially straightforward exercise to verify that the link length is invariant under the adjoint action of $SE(3)$. Indeed, clearly link length can be expressed in terms of 2-fold invariants:

$$a = \frac{I_{12}}{\sqrt{I_{11}I_{22} - I_{12}^2}}\left(\frac{\hat{I}_{11}}{I_{11}} + \frac{\hat{I}_{22}}{I_{22}}\right) - \frac{\hat{I}_{12}}{\sqrt{I_{11}I_{22} - I_{12}^2}} \tag{14}$$

The offset is the distance between the feet of successive common perpendiculars along the axis of the middle of three twists. In Figure (1b), the offset d_2 is the distance between $\mathbf{x}_2, \mathbf{x}_2'$, the feet of successive common perpendiculars from the axes of S_1, S_3 to the axis of S_2, so:

$$d_2 = \|\mathbf{x}_2 - \mathbf{x}_2'\| = \left\|\frac{\mathbf{v}_2 \times \boldsymbol{\omega}_2}{\boldsymbol{\omega}_2 \cdot \boldsymbol{\omega}_2} + t_2\boldsymbol{\omega}_2 - \frac{\mathbf{v}_2 \times \boldsymbol{\omega}_2}{\boldsymbol{\omega}_2 \cdot \boldsymbol{\omega}_2} + t_2'\boldsymbol{\omega}_2\right\| = |t_2 - t_2'|\|\boldsymbol{\omega}_2\|. \tag{15}$$

We must determine t_2, t_2'. If $\mathbf{x}_1, \mathbf{x}_2$ are the feet of the perpendicular between axes of S_1, S_2 then $\mathbf{x}_i = \dfrac{\mathbf{v}_i \times \boldsymbol{\omega}_i}{\boldsymbol{\omega}_i \cdot \boldsymbol{\omega}_i} + t_i\boldsymbol{\omega}_i$ for some t_i, $i = 1, 2$; the fact that $\mathbf{x}_2 - \mathbf{x}_1$ is perpendicular to axes of screws S_1 and S_2 gives us two equations and eliminating t_1 gives:

$$t_2 = \frac{((\mathbf{v}_2 \times \boldsymbol{\omega}_2) \cdot \boldsymbol{\omega}_1)(\boldsymbol{\omega}_1 \cdot \boldsymbol{\omega}_2) + ((\mathbf{v}_1 \times \boldsymbol{\omega}_1) \cdot \boldsymbol{\omega}_2)(\boldsymbol{\omega}_2 \cdot \boldsymbol{\omega}_2)}{\|\boldsymbol{\omega}_2\|^2\|\boldsymbol{\omega}_1 \times \boldsymbol{\omega}_2\|^2}. \tag{16}$$

A similar formula applies for t_2', from which we establish by (15), writing scalar triple products as 3×3 matrix determinants:

$$d_2 = \frac{\begin{pmatrix} \left(|\mathbf{v}_2\ \boldsymbol{\omega}_2\ \boldsymbol{\omega}_1|\,(\boldsymbol{\omega}_1 \cdot \boldsymbol{\omega}_2) + |\mathbf{v}_1\ \boldsymbol{\omega}_1\ \boldsymbol{\omega}_2|\,\|\boldsymbol{\omega}_2\|^2\right)\|\boldsymbol{\omega}_3 \times \boldsymbol{\omega}_2\|^2 \\ -\left(|\mathbf{v}_2\ \boldsymbol{\omega}_2\ \boldsymbol{\omega}_3|\,(\boldsymbol{\omega}_3 \cdot \boldsymbol{\omega}_2) + |\mathbf{v}_3\ \boldsymbol{\omega}_3\ \boldsymbol{\omega}_2|\,\|\boldsymbol{\omega}_2\|^2\right)\|\boldsymbol{\omega}_1 \times \boldsymbol{\omega}_2\|^2 \end{pmatrix}}{\|\boldsymbol{\omega}_2\|\,\|\boldsymbol{\omega}_3 \times \boldsymbol{\omega}_2\|^2\,\|\boldsymbol{\omega}_1 \times \boldsymbol{\omega}_2\|^2} \tag{17}$$

Once again it is possible to show that this is invariant under the action of $SE(3)$. In practice we used Maple show that it is invariant under the adjoint action of the Lie algebra on itself. However it is not immediately clear whether the numerator Ω_2 of this expression can be written in terms of the 14 invariants (10).

5 Algebro–Geometric Dual of Three Twists

Given a set $\mathscr{S}$ of three twists S_1, S_2, S_3, there are in fact three offsets determined by the three pairwise common perpendiculars, one associated with each twist. Recall from Sect. 2 that associated to each pair of twists S_i, S_j, their Lie bracket $[S_i, S_j]$ in the generic situation corresponds to a twist whose axis is the common perpendicular to the given pair. Thus we can associate with the given set a new dual 3-twist set $\mathscr{S}'$ consisting of the brackets $S_1' = [S_2, S_3]$, $S_2' = [S_3, S_1]$, $S_3' = [S_1, S_2]$, which has the property that the link lengths of $\mathscr{S}$ are the offsets of $\mathscr{S}'$ and vice versa. We can therefore determine an expression for the offset (17) in terms of the 3-fold invariants (10) by determining forms for the invariants of the dual set. By way of example, for $I_{11} = \boldsymbol{\omega}_1 \cdot \boldsymbol{\omega}_1$, we replace $\boldsymbol{\omega}_1$ by $\boldsymbol{\omega}_1' = \boldsymbol{\omega}_2 \times \boldsymbol{\omega}_3$ and then, using Lagrange's identity:

$$I_{11}' = (\boldsymbol{\omega}_2 \times \boldsymbol{\omega}_3)\cdot(\boldsymbol{\omega}_2 \times \boldsymbol{\omega}_3) = (\boldsymbol{\omega}_2 \cdot \boldsymbol{\omega}_2)(\boldsymbol{\omega}_3 \cdot \boldsymbol{\omega}_3) - (\boldsymbol{\omega}_2 \cdot \boldsymbol{\omega}_3)^2 = I_{22}I_{33} - I_{23}^2. \tag{18}$$

In the same way, we can determine expressions for the dual versions of all the invariants. For example, $I_{123}' = I_{123}^2$ and $\hat{I}_{123}' = 2I_{123}\hat{I}_{123}$.

We are now able to combine (14) with these expressions for the link lengths of $\mathscr{S}'$ to obtain an expression for the offset in terms of the basic invariants (10). Explicitly:

$$d_2 = \frac{\hat{I}_{123}I_{22}(I_{12}I_{23} - I_{13}I_{22}) + I_{123}\left(\hat{I}_{22}(I_{12}I_{23} + I_{13}I_{22}) + I_{22}(\hat{I}_{13}I_{22} - I_{12}\hat{I}_{23} - \hat{I}_{12}I_{23})\right)}{\sqrt{I_{22}}(I_{11}I_{22} - I_{12}^2)(I_{22}I_{33} - I_{23}^2)}. \tag{19}$$

Finally, note that the duality between $\mathscr{S}$ and $\mathscr{S}'$ is not an exact involution as, although the axes of the twists interchange, applying the process twice affects the pitches h_i'' of the double duals S_i'', $i = 1, 2, 3$. We obtain the following relation between the pitches: $h_i'' = \dfrac{\hat{I}_{123}}{I_{123}} + h_i$.

6 Conclusion

By applying the principle of transference to the well-known invariants of the rotation group, we have obtained basic invariants for sets of twists under Euclidean change of coordinates. Although the question has not been explored here, it can be shown (Daher 2013) that any m-fold polynomial invariant for $m \leq 3$ can be expressed as a rational function of these invariants. Ideally one would like to establish that they provide a generating set for the subalgebra of invariant polynomials. This appears to be a deep problem for $m \geq 3$. The virtue of these simple formulae is that they are defined for all sets of twists, unrestricted by special geometry, in contrast to Denavit–Hartenberg parameters. Indeed, we can find explicit algebraic expressions for the DH parameters in terms of them. A further issue is that our invariants are defined for a static set of twists. If the twists are regarded as defining joints in a serial chain, then the twists vary as the chain moves. So a second open question is whether the polynomials remain invariant under this intrinsic motion of the chain—DH parameters are invariant in this sense. A potential application for a set of polynomials, invariant in this full sense, would be to provide a firm basis for a classification of serial chains.

Acknowledgments The authors warmly acknowledge invaluable conversations with our colleague Dr Petros Hadjicostas (Victoria University of Wellington) and helpful advice from Professor Hanspeter Kraft (Basel University) and from Professor Manfred Husty (University Innsbruck) who drew our attention to the explicit formulation of invariants by Study.

References

Brockett R (1984) Robotic manipulators and the product of exponentials formula. In: Fuhrman P (ed) Proceedings of mathametical theory of networks systems. Springer, Berlin/Heidelberg, pp 120–129

Chevallier DP (1996) On the transference principle in kinematics: its various forms and limitations. Mech Mach Theory 31:57–76

Crook D (2009) Polynomial invariants of the Euclidean group action on multiple screws. Master's thesis, Victoria University of Wellington

Daher M (2013) Dual numbers and invariant theory of the Euclidean group. PhD thesis, Victoria University of Wellington (under examination)

Denavit J, Hartenberg RS (1955) A kinematic notation for lower pair mechanisms based on matrices. J Appl Mech 22:215–221

Donelan PS, Gibson CG (1991) First-order invariants of Euclidean motions. Acta Appl Math 24:233–251

Kotelnikov AP (1895) Screw calculus and some of its applications to geometry and mechanics. Annals of Imperial University of Kazan, Kazan

McCarthy JM (1986) Dual orthogonal matrices in manipulator kinematics. Int J Robot Res 5:45–51

Murray RM, Li Z, Shastry SS (1994) A mathematical introduction to robotic manipulation. CRC Press, Boca Raton

Panyushev DI (2007) Semi-direct products of lie algebras and their invariants. Publ Res Inst Math Sci 43:1199–1257

Perroud M (1983) The fundamental invariants of inhomogeneous classical groups. J Math Phys 24:1381–1391

Rico Martínez JM, Duffy J (1993) The principle of transference: history, statement and proof. Mech Mach Theory 28:165–177
Rooney J (1975) On the principle of transference. In: Fawcett JN (ed) Proceedings 4th world congress on the theory of machines and mechanisms, Institution of Mechanical Engineers, London, pp 10881092
Rooney J (2007) William Kingdon Clifford (1845–1879). In: Ceccarelli M (ed) Distinguished figures in mechanism and machine science: their contributions and legacies. History of mechanism and machine science, vol. 1. Springer, Dordrecht, Netherlands, pp 79116
Rosen J (1968) Construction of invariants for Lie algebras of inhomogeneous pseudo-orthogonal and pseudo-unitary groups. J Math Phys 9:1305–1307
Samuel AE, McAree P, Hunt KH (1991) Unifying screw geometry and matrix transformations. Int J Robot Res 10:454–470
Selig J (2005) Geometric fundamentals of robotics. Springer, New York
Study E (1903) Geometrie der Dynamen. Teubner, Leipzig
Takiff SJ (1972) Invariant polynomials on Lie algebras of inhomogeneous unitary and special orthogonal groups. Trans American Math Soc 170:221–230
Weyl H (1997) The classical groups: their invariants and representations. Princeton University Press, Princeton

Novel Quasi-Passive Knee Orthosis with Hybrid Joint Mechanism

Martin Huber, Matthew Eschbach, Horea Ilies and Kazem Kazerounian

Abstract Loss of personal mobility is a large and growing issue that can be caused by a variety of medical problems. Knee pain, muscle weakness, and their associated conditions are among the most common causes of impaired walking. Currently, conventional solutions include simple braces, canes, and medication. In more extreme cases, surgery is also routine. We propose a design for a novel quasi-passive orthotic knee brace which combines a smart support system employing magnetorheological (MR) fluid with a passive load reduction system. To work in concert with the walking motion of the knee, we have designed a four-bar linkage in conjunction with a compliant mechanism. The resulting orthotic knee brace will alleviate common symptoms related to knee pain and restore lost mobility, most notably in cases of osteoarthritis.

Keywords Quasi-passive · Knee orthosis · Osteoarthritis · Assistive robotics · Hybrid joint · Magnetorheological damper · Compliant mechanism

1 Introduction

Passive knee braces are an important tool in treating knee pain, particularly osteoarthritis (OA). Knee OA is characterized by degeneration of the articular cartilage and in severe cases collapse of bone at the femoral condyles (bony protrusions along which the knee joint articulates). Braces can help alleviate pain by shifting the load experienced by one of the two condyles to the other. Technology that addresses

M. Huber · M. Eschbach (✉) · H. Ilies · K. Kazerounian
University of Connecticut, Storrs, CT, USA
e-mail: matthew.eschbach@uconn.edu

M. Huber
e-mail: martin.huber@uconn.edu

H. Ilies
e-mail: ilies@engr.uconn.edu

K. Kazerounian
e-mail: kazem@engr.uconn.edu

A. Kecskeméthy and F. Geu Flores (eds.), *Interdisciplinary Applications of Kinematics*, Mechanisms and Machine Science 26, DOI 10.1007/978-3-319-10723-3_6

muscle weakness due to knee conditions while allowing full range of motion is not readily available. Some work has been done in the areas of fully active braces (Ferris et al. 2005), orthoses applying MR brakes to provide muscle support (Chen and Liao 2010; Lemaire et al. 2009), and stance-control orthoses involving springs (Irby et al. 2005). These devices either create motion through the use of actuators or supply a supportive torque about the knee joint to aid muscle function. Currently, no commercially available technology addresses both knee load and muscle activation reduction without the use of actuators or immobilizing the knee.

The design we propose is quasi-passive, rather than active. Accordingly, the system reacts to and supports motion, rather than creates it. By analyzing motion capture of the gait cycle, we have designed a brace which comprises three systems. First, a mechanism which follows the knee motion allows for a natural gait and a more comfortable brace. Second, a passive load acceptance system reduces the load experienced in the joint. Finally, a MR damper element provides supportive torque to the knee, reinforcing the quadriceps muscles at critical moments.

This system is capable of unloading both condyles while providing controllable muscle support, treating both muscle weakness, which is a cause and a symptom of knee pain, and knee pain itself. Additionally, the modular nature of the device and the controllability of the reinforcement system provide the design with a large amount of customization options. The degree of load reduction, the timing and magnitude of muscle support, and the knee follower mechanism all have the potential to be tailored to a given user.

2 Design Concept

2.1 The Gait Cycle

To better understand the purpose of this design, some background is required. From a standing position, the gait cycle begins with the initial contact of the lead limb with the ground and ends when that foot contacts the ground a second time. Once the initial contact occurs, the contact limb is said to be in its loading response phase, and is referred to as the stance limb. This phase is characterized by a loading response in the stance limb. The next phase, midstance, begins as the contralateral limb undergoes toe off. It ends as the bodys center of gravity is directly over the foot. During this interval it is crucial for the knee to recover from the initial contact shock in order to support the body as it moves forward over the stance limb. This is mainly conducted by the extensor (quadriceps) muscles in the thigh. Next is terminal stance that ends when the contralateral limb makes its initial contact. Afterwards, the limb enters swing, a phase in which all support is provided by the contralateral limb (Perry and Burnfield 2010; Winter 2005).

As can be guessed from the description of the gait cycle above, muscle support is critical during loading response and knee support is critical during the entire stance phase. Weakness in the extensor muscles can cause a significant decrease in the initial shock absorption capabilities of the knee. This sudden impact can cause further damage to the cartilage and pain in the knee joint. During stance, the knee can encounter loads higher than the total body weight (Morrison 1970). Reduction of these loads will lower pain and could provide a favorable environment for joint repair (Waller et al. 2011).

2.2 Design Objective and Process

The motion of the knee in the sagittal plane (dividing the left and right sides of the body) is not simple rotation; it is a mixture of sliding and rolling along a non-circular element (the femoral condyles) (Nuno and Ahmed 2001). It is commonly approximated as a pin joint, but this is a poor approximation—the center of rotation can move several centimeters during knee flexion (Smidt 1973). Our first goal is to find a mechanism to connect the thigh and calf pieces of the orthosis. It must be able to replicate the characteristic knee rotation. Due to the nature of this motion, possibilities involve non-circular gears, cam systems, or four-bar linkages. These all prescribe a precise motion, however, and may not allow for inconsistencies in gait or elasticity inside the joint.

By allowing relative motion in one degree of freedom, namely allowing translation vertically along the calf, we found that the coupler of a 4R four-bar linkage could effectively maintain alignment with the calf's body vector. This relative motion creates the desired leniency, as well as affords us the opportunity to introduce a resistive element inside the gap, as will be discussed shortly. In this configuration, the 4R four-bar linkage connects the thigh and coupler pieces of the orthosis. This is then paired with a prismatic joint between the coupler link and the calf portion of the brace. Both the calf and thigh portions of the brace are fixed to their respective leg segments.

The first support element is the load reduction system. The coupler-calf relationship explained above leads to a straightforward concept—placing an elastic element between these two links. The requirements are twofold: this element must be active only during the stance phase, and it must take some of the knee's load, which will be transmitted around the knee through the four bar linkage. To meet the first criterion, we fix the elastic element to the calf link, and design the motion of four bar such that the coupler only compresses this element when the leg is near extension. The second is slightly more involved. We have a desired load reduction and a coupler-calf gap size, both dependent on calf angle, causing our desired elastic element to have a nonlinear force-displacement relationship. This can be achieved by a cam system, which would modify the gap size in such a way to allow a linear spring to be used, or by compliant mechanisms, for example a bistable beam combined with a linear spring (Chen and Lan 2012). A leaf spring design possessing arc geometry was chosen. It can be produced simply with 3D printing technology, can be designed for a given force-displacement curve, and does not require the introduction of moving parts. These properties allow it to be easily designed for a custom brace.

Muscle reinforcement must come in the form of supportive torque during loading response. Because we desire support at this point and unimpeded leg motion elsewhere, it must be able to either be controlled or switched on and off. A mechanical release/clutch system based on leg angle encounters problems due the fact that the same angle can be seen in both swing and stance. Also, it introduces a failure mode that could result in an on-state rather than the default off-state, leaving the user unable to freely swing the leg. An acceptable solution is an electronically controllable device, namely one involving MR fluid. As previously mentioned, rotary MR brakes have been investigated for this purpose. Our design does not possess a stationary center of rotation, limiting the viability of a fixed axis rotary brake. We can, however, implement a linear MR damper and use the linkage motion to convert damping force into a useful torque. A depiction of all three systems, which have been separated and exaggerated for clarity, can be seen above in Fig. 1.

Because this device will be worn, size and weight are restricted. In literature and industry, common weights of passive braces are as little as 0.5 kg. Being designed for those who already experience leg weakness, the weight will be minimized. In addition, power consumption by the electromagnetic system must allow for continuous daily use.

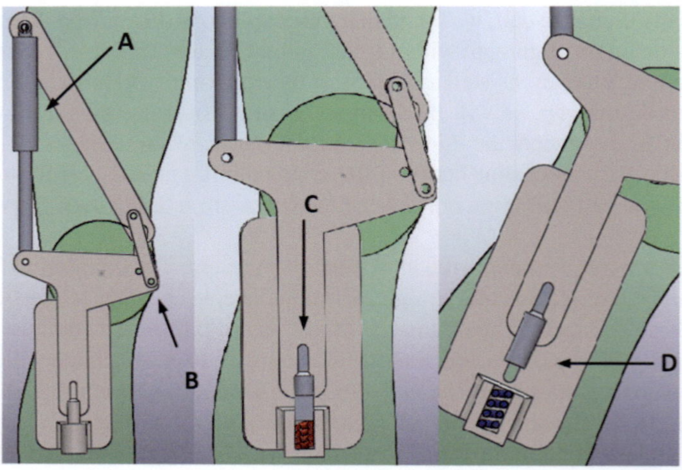

Fig. 1 Design concept *A* Damper, *B* Four bar linkage, *C* Compressed spring during extension, *D* Uncompressed spring and prismatic joint during flexion

3 Design Components

3.1 Mechanism

Motion capture of a subject's gait is used to define the motion for which the orthosis is designed (Carnegie Mellon University 2014). After some processing, the motion of the calf is described with respect to a vertical and stationary thigh by a series of displacement matrices. The 4R linkage is designed with intentional relative motion between the target (calf) and the coupler link. This motion is restricted to being along the body vector of the calf, and is responsible for the compression of the spring. Because the spring will be designed for this motion, there is a range of acceptable relative motion paths between the two. As a result, the four-bar synthesis problem becomes one of rigid body guidance through a series of rays rather than poses; i.e. the coupler body vector must lie along the ray of the calf body vector. Because the spring is designed to engage at calf angles less than approximately 20 degrees from vertical (associated with stance phase), the gap size at all angles greater than 20 degrees must be larger than that at 20. Moreover, angles lower than 20 must have a smaller gap size.

This can be posed as a three position synthesis problem with only the fixed pivot locations and the rotation portions of the displacement matrices specified. These conditions can be seen in Fig. 2. Using the constant length constraints of the input, coupler, and output links (Eqs. 1, 2, 3), and constraining the x-y displacement to bring the coupler to the specified ray (Eq. 4), a system of six nonlinear equations and six unknowns emerges.

$$([D_i]a_1 - [D_i]b_1)^T([D_i]a_1 - [D_i]b_1) = (a_1 - b_1)^T(a_1 - b_1) \quad (1)$$

$$([D_i]a_1 - a_0)^T([D_i]a_1 - a_0) = (a_1 - a_0)^T(a_1 - a_0) \quad (2)$$

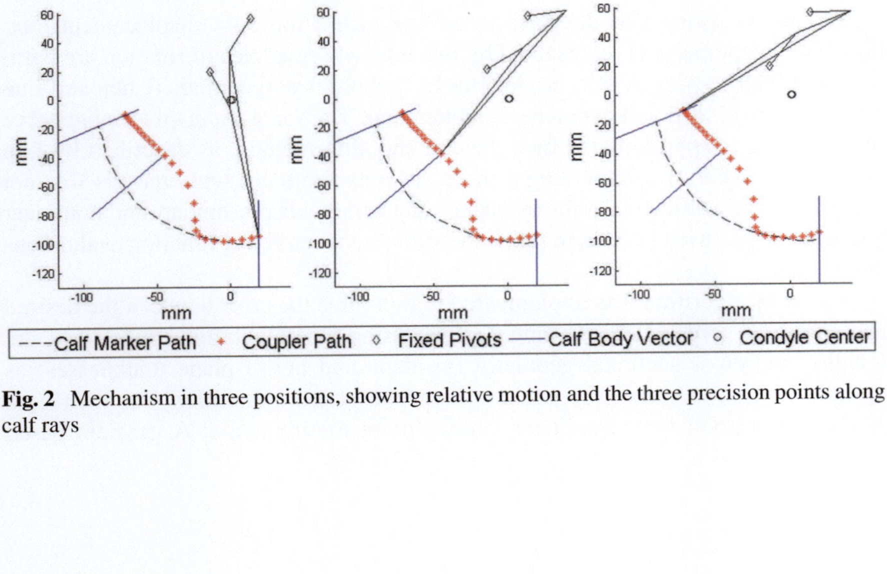

Fig. 2 Mechanism in three positions, showing relative motion and the three precision points along calf rays

$$([D_i]b_1 - b_0)^T([D_i]b_1 - b_0) = (b_1 - b_0)^T(b_1 - b_0) \tag{3}$$

$$\begin{bmatrix} \delta x \\ \delta y \end{bmatrix}_i = P_i + s_i u_i - [R_i]P_0 \tag{4}$$

Where i goes from 1 to 2, a_0 and b_0 are the fixed pivot locations, a_1 and b_1 are the moving pivots, P_i is the calf marker at ray i, which possesses the unit vector u_i. s_i is the distance along the ray from P_i. δx and δy are the translations at displacement i.

A least squares minimization was used to solve the above system and perform a direct search for fixed pivot locations, which were then refined by local polling. The designs were evaluated for the relative motion within limits, the degree of correspondence to the angle trajectory, and the vectors of force transmission to the ground (thigh) during stance.

This optimization system was able to produce a coupler path that tracked the calf body vector almost exactly—the average deviation from the leg angle was less than one degree. Also, its size and force transfer characteristics were satisfactory. The link lengths are acceptable, and the resultant vector of force transfer from coupler to ground is close to vertical during stance phase.

3.2 Custom Spring Design

Force plate and motion capture data were used to determine the load the system is required to absorb (Winter 2005). A percentage of the maximum natural load was defined (in this case 60) as the new maximum the system would allow the knee to experience. This leads to a flat knee load vs. gait percent curve, which is physically unobtainable due to the fact that the same angle is repeated during stance phase with different spring force requirements. To produce the target points, the maximum load at select angles was taken, leading to a relationship between spring force and spring displacement.

A titanium spring was designed to realize such a force vs. displacement relationship. Its geometry is represented by two arcs whose angles of rotation are parametrized with respect to their arc lengths by 2rd order polynomials (Chen and Lan 2012). Its cross sectional geometry is rectangular. The force vs. displacement curve of the spring is approximated by a general shooting method, as described by Lan and Cheng (2008). Implementing a more accurate finite element analysis was not computationally efficient for the synthesis, due to large displacements and nonlinear geometries. It is used later on to refine the structure when fewer function evaluations are needed.

A genetic algorithm was implemented to minimize the error between the desired and simulated force vs. displacement curves using the polynomial coefficients, arc lengths, and cross sectional geometry (in-plane and out-of-plane thicknesses) as optimization parameters. The solution from the genetic algorithm was then used as the initial point for a nonlinear interior point minimization. A gradient based

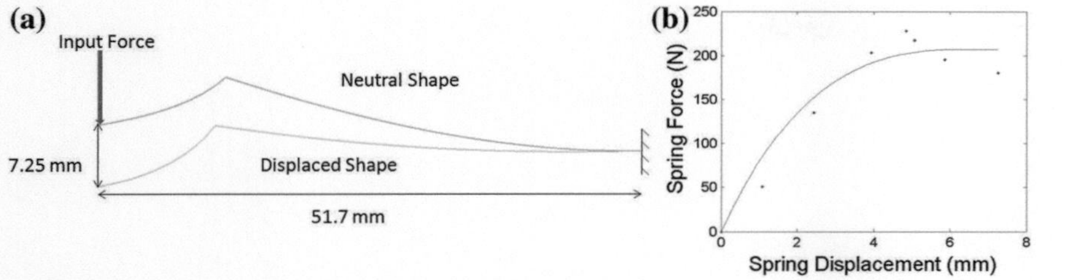

Fig. 3 **a** Spring, neutral and compressed, **b** Ideal (*points*) and designed (*solid*) force versus displacement *curve*

optimization evaluated through finite element analysis was used to obtain a final spring solution, an outcome of which can be seen in Fig. 3 (Jutte and Kota 2008).

3.3 Damper

The function of the damper element is to apply a moment resistive to knee flexion during portions of the stance phase. As described in Sect. 2.2, magnetorheological fluid was implemented. Through finite element magnetics evaluation in combination with a simple flow-mode Bingham plastic model, standard piston dampers were designed (Zhu et al. 2012). Their placement between either the coupler-thigh links or the coupler-calf links creates the desired supportive moment.

4 Validation

In order to confirm that the basic design configuration behaves as expected, a kinetostatic link/segment model of the gait cycle was developed from sagittal plane motion capture and force plate data. This is a common method used in kinesiology and gait analysis to evaluate joint torques and forces (Alkjaer et al. 2001; Winter 2005). From this model, a progression of the load experienced by the knee during normal walking is calculated. The design components are then added to the free body diagram of the knee, and the orthotic system superimposed onto the model. By comparing the load and moment of an unaided knee to an aided knee, the validity of the concept can be evaluated. Figure 4 shows results from this simulation: an expected reduction in both the load experienced (a) and the moment required (b). While this may show us that the theoretical kinematics of the design function as intended, a more detailed picture is still needed.

In addition, a prototype will be fabricated and tested, first in a bench testing rig to ensure proper function, then in live trials. These will be conducted with the help of force plate and motion capture systems to evaluate gait changes due to the device. We

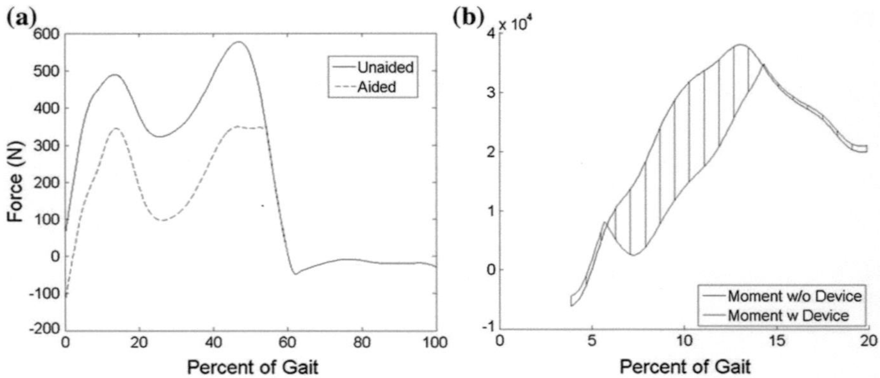

Fig. 4 **a** Load reduction, **b** Moment reduction

also need subjective data, and will obtain it through standard surveys used in physical therapy, such as the Knee Injury and Osteoarthritis Outcome Score (KOOS), which measures the patients' impressions of changes in their pain, ease of motion, and knee function Knee injury and Osteoarthritis Outcome Score (2014).

5 Conclusions and Future Efforts

Our initial simulation suggests that our device functions as intended, though it remains to be seen whether the expected pain relief and ease of stance outcomes have been realized. To that end, we plan to move forward with prototyping and testing the system.

Further development of the device will include refining a sensor strategy to detect which phase of gait the limb finds itself in, and a corresponding response strategy to apply the correct resistive torque. At the moment, the system has been designed and simulated in an environment that involves steady gait on flat ground. The system must be expanded to recognize several other common states, most importantly stair descent and ascent, standing in place, and turning. In many of these cases, the support necessary during stance will be identical or similar to flat gait, with new criteria during flexion and extension. Some design changes may be necessary, most probable is the moment arm of the damper with respect to the rotation of the calf; it has been optimized only for stance phase during normal gait. Additional work must then be done to determine proper reactions to unusual or uneven terrain and changing gait patterns, particularly stride length and speed.

This design is based on the motion capture of an individual's gait, and has a highly modular nature. Accordingly, it is possible to individualize the system components to a specific user. Motion characteristics, load reduction requirements, and muscle support patterns can all be custom designed. To facilitate this customization, we plan to investigate the effects of variations on the design parameters. This will lead to associating particular gait parameters with areas of the design space, ultimately creating a more efficient design process. The goal of this work would be an automated system which receives gait motion capture data and produces an acceptable orthosis design.

References

Alkjaer T, Simonsen EB, Dyhre-Poulsen P (2001) Comparison of inverse dynamics calculated by two- and three-dimensional models during walking. Gait Posture 13(2):73–77 (Apr. 2001. 12:257–271)

Carnegie Mellon University—CMU Graphics Lab—motion capture library. http://mocap.cs.cmu.edu/. Accessed 17 Oct 2012

Chen JZ, Liao WH (2010) Design, testing and control of a magnetorheological actuator for assistive knee braces. Smart Mater Struct 19(3):035029

Chen Y-H, Lan C-C (2012) An adjustable constant-force mechanism for adaptive end-effector operations. J Mech Des 134(3):031005–031005

Ferris DP, Czerniecki JM, Hannaford B (2005) An ankle-foot orthosis powered by artificial pneumatic muscles. J Appl Biomech 21(2):189–197

Irby SE, Bernhardt KA, Kaufman KR (2005) Gait of stance control orthosis users: the dynamic knee brace system. Prosthet Orthot Int 29(3):269–282

Jutte CV, Kota S (2008) Design of nonlinear Springs for prescribed load-displacement functions. J Mech Des 130(8):081403–081403

Knee injury and Osteoarthritis Outcome Score, Ewa roos. http://www.koos.nu/. Accessed 09 Aug 2013

Lan C-C, Cheng Y-J (2008) Distributed shape optimization of compliant mechanisms using intrinsic functions. J Mech Des 130(7):072304–072304

Lemaire ED, Goudreau L, Yakimovich T, Kofman J (2009) Angular-velocity control approach for stance-control orthoses. IEEE Trans Neural Syst Rehabil Eng 17(5):497–503

Morrison JB (1970) The mechanics of the knee joint in relation to normal walking. J Biomech 3(1):51–61

Nuno N, Ahmed AM (2001) Sagittal profile of the femoral condyles and its application to femorotibial contact analysis. J Biomech Eng 123(1):18–26

Perry J, Burnfield JM (2010) Gait analysis: normal and pathological function, 2nd edn. SLACK Incorporated, Thorofare

Smidt GL (1973) Biomechanical analysis of knee flexion and extension. J Biomech 6(1):79–92

Waller C, Hayes D, Block JE, London NJ (2011) Unload it: the key to the treatment of knee osteoarthritis. Knee Surg Sports Traumatol Arthrosc 19(11):1823–1829

Winter DA (2005) Biomechanics and motor control of human movement, 3rd edn. Wiley, Hoboken

Zhu X, Jing X, Cheng L (2012) Magnetorheological fluid dampers: A review on structure design and analysis. J Intell Mater Syst Struct

Four-Position Synthesis of Origami-Evolved, Spherically Constrained Planar RR Chains

Kassim Abdul-Sater, Franz Irlinger and Tim C. Lueth

Abstract In this paper we present a dimensional four-position synthesis procedure for an overconstrained 1-DOF spatial linkage that we call the origami-evolved, spherically constrained planar RR chain. The structure is found as a mechanism equivalent of a part of the Miura-ori folding linkage. Studying the geometry of this mechanism equivalent it turns out that it only corresponds to two spherical four-bar linkages that are coupled in a special overconstraining manner. Even though the special characteristic of the coupling is necessary to preserve 1-DOF mobility, there is particular freedom left for the design and synthesis of the linkage. On the one hand it is allowed to use a spherical or either a spatial RR chain when constructing an origami-evolved, spherically constrained RR chain. On the other hand the synthesis of spherical RR dyads becomes available for the design of the spherical wrists.

Keywords Rigid Miura-ori pattern · Origami mechanism equivalent · Planar and spherical four-position synthesis

1 Introduction

Apart from origami art objects, some so called rigid origami folding patterns exist, which remain moveable, performing 1-DOF motion, even if paper and creases are replaced by rigid panels and revolute joints. These foldings represent overconstrained linkages, which are moveable because of highly symmetric crease patterns.

A common example of a rigid origami is the Miura-ori pattern (Miura 1985, Fig. 1a), also used to deploy solid solar panels in space. It is built from a basic

K. Abdul-Sater (✉) · F. Irlinger · T. C. Lueth
Institute of Micro Technology and Medical Device Technology, Faculty of Mechanical Engineering, Technische Universität München, Garching, Germany
e-mail: kassim.abdul-sater@tum.de

F. Irlinger
e-mail: irlinger@tum.de

T. C. Lueth
e-mail: tim.lueth@tum.de

A. Kecskeméthy and F. Geu Flores (eds.), *Interdisciplinary Applications of Kinematics*, Mechanisms and Machine Science 26, DOI 10.1007/978-3-319-10723-3_7

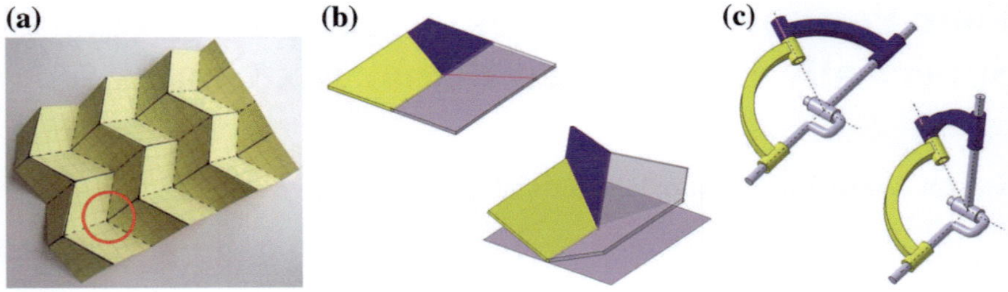

Fig. 1 **a** The Miura-ori folding pattern, **b** CAD model of the Miura single vertex pattern (MSV), **c** Mechanism equivalent of the rigid MSV-pattern

pattern that we call the *Miura single-vertex (MSV)*, which consists of a symmetrical arrangement of four intersecting creases and four plane rigid panels. These together form a spherical symmetrical four-bar mechanism (Fig. 1b).

While Tachi (2009) analyzed the Miura pattern and generalized its rigid foldability for further quadrilateral origami meshes we restrict ourself to the building block, which was the MSV. Observing the plane-symmetric movement shown in Fig. 1b, it is easy to see that one half of the linkage can be replaced by two new links that play the role of the line of symmetry in the MSV (Fig. 1c). This is because the new links are hinged in a way, that the resulting four-bar performs movement identical to that of the original linkage. An interesting aspect is that the linkage from Fig. 1c is also found as a substructure of a mechanism equivalent of a folded carton box by Wei and Dai (2009).

If we couple two of these linkages as shown in Fig. 2a another *spatial* 1-DOF mechanism is obtained because the links forming the moving vertex perform spatial movement. We call this linkage the *Miura double vertex mechanism equivalent (MDV)*, and one can see that it includes a planar RR chain. The linkage consists of $g = 7$ revolute joints, each providing the freedom $b_i = 1$, and $n = 6$ links: K', K'' and K''' (forming the RR chain) and *K1*, *K2* and *K3* (forming the spherical wrists). Applying the mobility formula for spatial mechanisms yields

$$f = 6\,(n - 1 - g) + \sum_{i=1}^{g} b_i = 6\,(6 - 1 - 7) + 7 = -5, \qquad (1)$$

which shows that the MDV is an overconstrained linkage. Never-the-less, 1-DOF mobility of the structure is ensured because the link K'' connecting the revolute axes of the RR chain provides the hinge of $K2$ which preserves each of the vertices formed by the spherical 4R linkages. This is equivalent to saying that the six common normal distances (six geometric conditions, see also Abdul-Sater et al. 2013) among this hinge and each of the remaining ones remain zero when the linkage moves.

However, this means that the MDV can be transformed into various further 1-DOF mechanisms. On the one hand a spherical or spatial RR chain can be used instead of a planar one (Fig. 2b). On the other hand each vertex can be designed independently

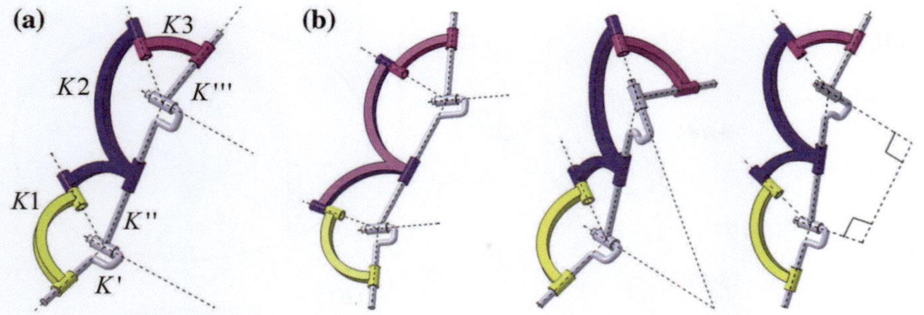

Fig. 2 **a** The Miura double vertex mechanism equivalent (MDV), **b** Examples of allowed variations of shape of an MDV mechanism equivalent: the origami-evolved, spherically constrained planar, spherical and spatial RR chain

in shape. We call the family of these mechanisms origami-guided RR chains (see also Abdul-Sater et al. 2013) or now origami-evolved, spherically constrained RR chains.

The mobility of an origami-guided RR chain is ensured, as long as the line connecting the revolute axes of the RR chain represents the hinge axis between K'' and $K2$ and coincides with each of the vertices formed by the remaining hinges. This leads to the fact that particular freedom exists for the synthesis of the family of these linkages.

In (Abdul-Sater et al. 2013) we applied a two-configuration synthesis method to the origami-guided chains. This allowed it to constrain consecutive links of the structure, such that these can reach two given spatial positions. Extending our research on these linkages, we now present a synthesis procedure, which systematically applies the well-developed four-position synthesis of planar and spherical RR chains based on geometric constraints to obtain an origami-guided *planar* RR chain. We restrict ourself to four positions to obtain a larger number of linkage solutions as would be obtained for the maximum number for planar and spherical RR synthesis which is five (see e.g. McCarthy and Soh 2010).

2 Preliminaries

To accomplish the synthesis of the linkage we introduce the following coordinate frames (see Fig. 3): First a fixed world frame W is defined. To describe the movement of the links from Fig. 2a, we introduce the following frames: Another fixed frame B', similarly oriented as W. Two moving frames B'' and B''' for the RR chain, and three further moving frames $B1$, $B2$ and $B3$ for the remaining links.

B', B'', $B1$ and $B2$ shall be located at the fixed vertex, B''' and $B3$ are located at the moving vertex. B' and B'' always have their z-axis aligned with the line $\mathsf{L}_{B'|B''}$. $B2$ performs spherical motion around the origin of B' which is a composition of

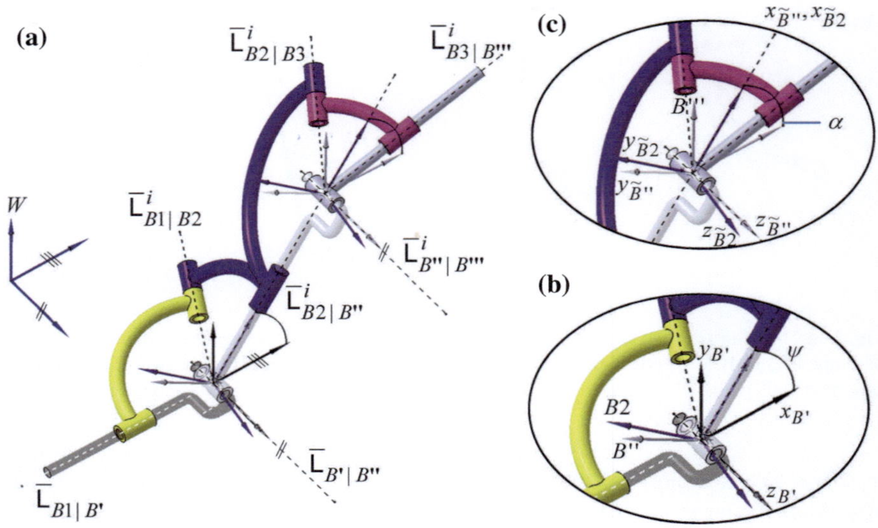

Fig. 3 **a** Nomenclature to describe the hinges of origami-evolved, spherically constrained planar RR chains **b** Frames located at the fixed vertex **c** Frames located at the moving vertex

the movement of B'' and a rotation around the hinge $\bar{L}^i_{B2|B''}$ (Fig. 3b). B''' has its z-axis aligned with the axis of the moving R joint of the RR chain and therefore performs planar movement in the xy-plane of B' (Fig. 3c). Rigidly attached to B''' but differently located we define $\tilde{B}'''$ (not shown in Fig. 3) which will be needed to pre-define the task positions of the end-link of the RR chain. We define two further frames $\tilde{B}''$ and $\tilde{B}2$, needed for the synthesis of hinges at the second vertex. These frames are rigidly attached to the links with corresponding frames B'' and $B2$ and are located at the moving vertex (Fig. 3c).

The lines associated with the different revolute axes are denoted using a nomenclature that expresses the couplings of the links or frames. The right upper index '*i*' denotes a configuration of the linkage and is needed to describe the four-position problem. A line coupling two frames Ba and Bb shall be described using *Plücker coordinates* (see e.g. McCarthy and Soh 2010): $\bar{L}^i_{Ba|Bb} = (\mathbf{d}^i, \mathbf{p}^i \times \mathbf{d}^i)_{Ba|Bb}$. Herein $\mathbf{d}^i_{Ba|Bb}$ is the direction of the line, measured in W. $\mathbf{p}^i_{Ba|Bb} \times \mathbf{d}^i_{Ba|Bb}$ defines the moment about the origin of W, where $\mathbf{p}^i_{Ba|Bb}$ defines the location of the line.

3 Four-Position Synthesis Procedure

$\bar{L}_{B2|B''}$ ensures the mobility of the origami-evolved, spherically constrained RR chains since it represents the revolute axis of the link $K2$ (Fig. 2). Hence, since this axis can be defined by the revolute axes of the RR chain, the first step in the synthesis

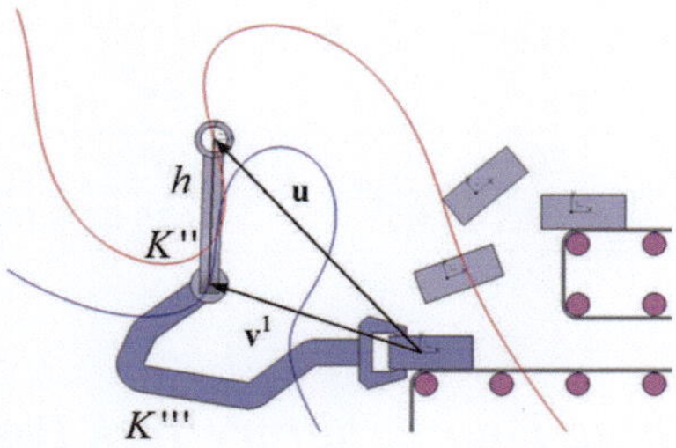

Fig. 4 The center-axis and the center-point *curve* determined from the RR synthesis equations and the four poses of $\tilde{B}'''$

Fig. 5 Two center-axis and circling-axis cones, defining the spherical RR dyads at the fixed and moving pivot of the planar RR chain

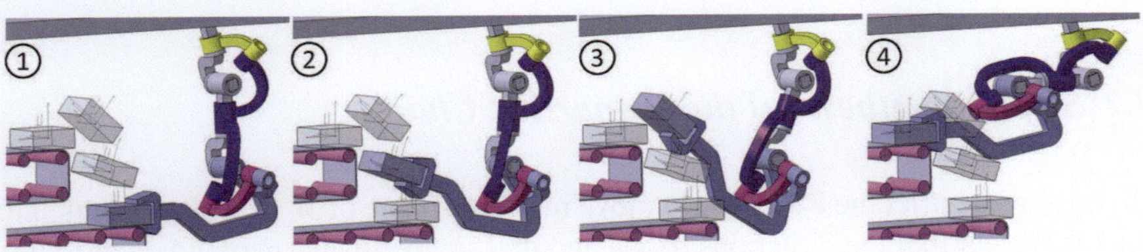

Fig. 6 CAD design of the linkage reaching the four positions defining a pick-and-place task

procedure is to synthesize the planar RR chain for four pre-defined positions in the world frame W.

Figures 4 and 5 depict the different steps of the synthesis procedure using the example of a planar pick-and-place task. Figure 6 shows a detailed CAD design of the origami-evolved, spherically constrained planar RR chain reaching the four pre-defined positions.

Table 1 Arbitrarily chosen task position for the frame B'''

	$\tilde{B'''}^1$	$\tilde{B'''}^2$	$\tilde{B'''}^3$	$\tilde{B'''}^4$
ϕ^i	0°	20°	40°	0°
$\mathbf{t}^i$	$(0\ 0\ 0)^T$	$(7.5\ 20\ 0)^T$	$(15\ 42.5\ 0)^T$	$(40\ 37.5\ 0)^T$

3.1 Step 1: Pre-defining Four Arbitrary Task Positions

In order to pre-define the four planar positions of the end-link K''' or $\tilde{B}'''$ of the RR chain we use the *displacement equation*

$$\mathbf{x}^i = \mathbf{R}^i_{\tilde{B}'''}\xi + \mathbf{t}^i_{\tilde{B}'''}, \qquad i = 1, \ldots, 4, \tag{2}$$

where $\mathbf{t}^i_{\tilde{B}'''}$ is a 3×1 translation vector and $\mathbf{R}^i_{\tilde{B}'''}$ is a proper 3×3 rotation matrix, both together defining a planar displacement in the xy-plane of W (see Table 1). Solving the first equation for ξ and substituting this into the remaining equations yields the *relative displacements* which allow it to express $\mathbf{x}^i$ by $\mathbf{x}^1$:

$$\mathbf{x}^i = \underbrace{\mathbf{R}^i_{\tilde{B}'''}(\mathbf{R}^1_{\tilde{B}'''})^T}_{\mathbf{R}^{1i}_{\tilde{B}'''}}\mathbf{x}^1 + \underbrace{\mathbf{t}^i_{\tilde{B}'''} - \mathbf{R}^i_{\tilde{B}'''}(\mathbf{R}^1_{\tilde{B}'''})^T\mathbf{t}^1_{\tilde{B}'''}}_{\mathbf{t}^{1i}_{\tilde{B}'''}}, \qquad i = 2, \ldots, 4. \tag{3}$$

The benefit of this expression for the synthesis of our linkage is that in the following we will be able to determine its hinges in a reference position 1.

3.2 Step 2: Synthesis of the Planar RR Chain

Because we restrict the RR chain to move in the xy-plane of W or a plane parallel to that, it is clear that the directions $\mathbf{d}_{B'|B''}$ and $\mathbf{d}^1_{B''|B'''}$ are always perpendicular to this plane. Hence we are only interested in the locations $\mathbf{u} := \mathbf{p}_{B'|B''}$ and $\mathbf{v}^1 := \mathbf{p}^1_{B''|B'''}$, such that the resulting planar RR chain can reach the task positions.

We follow McCarthy et al. (2010) who use the algebraic expression

$$(\mathbf{v}^i - \mathbf{u})^T(\mathbf{v}^i - \mathbf{u}) = h^2, \quad i = 1, \ldots, n \tag{4}$$

of the constant distance h between the locations of the hinges to derive the *design equations* of a planar RR chain. Note that $\mathbf{u}$ and $\mathbf{v}^1$ shall be defined using the same z-coordinates in W. After setting $n = 4$, subtracting the first equation of Eq. (4), substituting $\mathbf{v}^i = \mathbf{R}^{1i}_{\tilde{B}'''}\mathbf{v}^1 + \mathbf{t}^{1i}_{\tilde{B}'''}$ and some rearranging, the set of three scalar bilinear design equations is obtained as:

$$\mathbf{u}^T(\mathbf{E}-\mathbf{R}^{1i}_{\tilde{B}'''})\mathbf{v}^1+(\mathbf{t}^{1i}_{\tilde{B}'''})^T\mathbf{R}^{1i}_{\tilde{B}'''}\mathbf{v}^1-(\mathbf{t}^{1i}_{\tilde{B}'''})^T\mathbf{u}+\frac{1}{2}(\mathbf{t}^{1i}_{\tilde{B}'''})^T\mathbf{t}^{1i}_{\tilde{B}'''}=0,\quad i=2,\ldots,4, \quad (5)$$

where $\mathbf{E}$ is the identity matrix. Equation (5) can be solved to obtain the well-known center-point and circle-point curve that represent all the valid *locations* for the hinges of the planar RR chain. We selected the center and circle point in the xy-plane of W and designed the RR chain, shown in Fig. 4.

3.3 Step 3: Pre-defining Four Positions of the Link K2

Recall now from Sect. 2, Fig. 3 that the link $K2$ or frame $B2$ performs spherical motion around the origin of the fixed frame B' which is a composition of the z-rotation of B'' and a x-rotation around the hinge $\bar{\mathsf{L}}_{B2|B''}$:

$$\mathbf{R}^i_{B2}=\mathbf{Z}(\psi^i)\mathbf{X}(\vartheta^i)=\begin{pmatrix}\cos\psi & -\cos\vartheta\sin\psi & \sin\psi\sin\vartheta\\ \sin\psi & \cos\psi\cos\vartheta & -\cos\psi\sin\vartheta\\ 0 & \sin\vartheta & \cos\vartheta\end{pmatrix}^i \quad (6)$$

Since the movement of B'' is already determined by the synthesis of the RR chain, the angles $\psi^i,\ i=1,\ldots,4$ exist and are easily determined by the following inverse kinematics calculations:

$$\psi^i=\arctan\left(\frac{v^i_y-u_y}{v^i_x-u_x}\right),\quad i=1,\ldots,4, \quad (7)$$

For the second part $\mathbf{X}(\vartheta^i)$ of the composition we are free to select four arbitray angles which is because we do not impose any constraint to the RR chain. Table 2 lists the values for ψ and ϑ that were calculated and chosen for the design example.

Table 2 Angular position data for the frame *B*2

Position '*i*'	1	2	3	4
ψ^i	268.122°	274.889°	292.534°	358.172°
ϑ^i	0°	20°	40°	120°

3.4 Step 4: Adding Spherical RR Chains

To constrain the movement of the planar RR chain, we use the synthesis of spherical RR dyads based on geometric constraints, which yields the directions $\mathbf{g} := \mathbf{d}_{B1|B'}$, $\mathbf{h}^1 := \mathbf{d}^1_{B1|B2}$, $\mathbf{r}^1 := \mathbf{d}^1_{B3|B'''}$ and $\mathbf{s}^1 := \mathbf{d}^1_{B2|B3}$ of the remaining axes of the linkage. However, to achieve this goal it is convenient to state the synthesis equations with respect to the frames B' and $\tilde{B}''$ (see Fig. 3), located at fixed and moving vertex that were previously determined as $\mathbf{u} := \mathbf{p}_{B'|B''}$ and $\mathbf{v}^i := \mathbf{p}^i_{B''|B'''}$.

McCarthy et al. (2010) show how to manipulate the algebraic expression of the constant angle between the two directions $\mathbf{a}^i = \mathbf{R}^{1i}_{Ba}\mathbf{a}^1$ and $\mathbf{b}^i = \mathbf{R}^{1i}_{Bb}\mathbf{b}^1$ of a general spherical RR dyad to derive the bilinear *design equations* for a four-position task:

$$\mathbf{a}^{1^T}(\mathbf{E} - (\mathbf{R}^{1i}_{Ba})^T\mathbf{R}^{1i}_{Bb})\mathbf{b}^1 = 0, \quad i = 2, \ldots, 4. \tag{8}$$

Note that the frames Ba and Bb need to correspond to non-consecutive links. The solution of this set of three bilinear equations yields the well-known *center-axis* and *circling-axis cone* that represent all the valid *directions* for the hinges.

We state (8) for the hinges at the fixed vertex in frame B'. Since B' and W have the same orientation, the vectors $\mathbf{g}$ and $\mathbf{h}^1$ have the same coordinates in B' and W and we can state (8) as

$$\mathbf{g}^T(\mathbf{E} - \mathbf{R}^{1i}_{B2})\mathbf{h}^1 = 0, \quad i = 2, \ldots, 4. \tag{9}$$

Note that B' is fixed, which yields $\mathbf{R}^{1i}_{B'} = \mathbf{E}$.

We compute the coordinates of $\mathbf{r}^1$ and $\mathbf{s}^1$ in $\tilde{B}''$, denoted as ${}^{\tilde{B}''}\mathbf{r}^1$ and ${}^{\tilde{B}''}\mathbf{s}^1$. This is because the frames B''' and $\tilde{B}2$, which will be connected by the dyad, perfom pure rotations about the origin of $\tilde{B}''$. These rotations shall be denoted as ${}^{\tilde{B}''}\mathbf{R}^{1i}_{B'''}$ and ${}^{\tilde{B}''}\mathbf{R}^{1i}_{\tilde{B}2}$ and allow us to write Eq. (8) as:

$${}^{\tilde{B}''}\mathbf{r}^{1^T}(\mathbf{E} - ({}^{\tilde{B}''}\mathbf{R}^{1i}_{B'''})^T\,{}^{\tilde{B}''}\mathbf{R}^{1i}_{\tilde{B}2})\,{}^{\tilde{B}''}\mathbf{s}^1 = 0, \quad i = 2, \ldots, 4. \tag{10}$$

The movement of $\tilde{B}2$, seen in $\tilde{B}''$, is given as a x-rotation by the predefined angles ϑ^i from Table 2. That of B''' is a z-rotation by an angle 'α' measured among the xz-planes of $\tilde{B}''$ and B''' (Fig. 3). This is easily obtained from the inverse kinematics calculation $\alpha^i = \phi^i - \psi^i$, $i = 1, \ldots, 4$ (see also Tables 1 and 2).

Figure 5 shows the two different center-axis (red) and circling-axis cones (blue) obtained from a solution of Eqs. (9) and (10). We selected the axes $\mathbf{g}$ and $\mathbf{h}^1$ at the fixed vertex and $\mathbf{r}^1$ and $\mathbf{s}^1$ at the moving vertex and created a detailed CAD design of the complete linkage. Figure 6 shows the design which is free from collisions and which can reach the four task positions by only actuating the link K''. The yellow- and purple-colored links define the spherical RR dyads.

4 Conclusions

The usual way to constrain the movement of a planar RR chain for a given finite position task is to perform synthesis step 2 (Sect. 3.2) and select two of these chains at the center point and circle point curve to obtain a four-bar. Compared to this classic approach, we showed how to constrain such an RR chain for a given task by designing two spherical wrists with respect to a single condition. This yields two further design steps in the dimensional synthesis but provided additional free selectable parameters (angles). The design approach may always yield a slim linkage design and is also suitable to constrain spherical and spatial RR chains.

References

Abdul-Sater K, Irlinger F, Lueth TC (2013) Two-configuration synthesis of origami-guided planar. In: McCarthy JM (ed) Spherical and spatial revolute-revolute chains. ASME J Mech Robot 5(3):10

McCarthy JM, Soh GS (2010) Geometric design of linkages, 2nd edn. Interdisciplinary applied mathematics. Springer, New York

Miura K (1985) Method of packaging and deployment of large membranes in space. Science report No. 618, Institute of space and astronautical science

Tachi T (2009) Generalization of rigid foldable quadrilateral mesh origami. In: Domingo A, Lazaro C (eds) Proceedings of the international association for shell and spatial structures

Wei G, Dai JS (2009) Geometry and kinematic analysis of an origami-evolved mechanism based on artmimetics. In: Dai JS, Zoppi M, Kong X (eds) ASME/IFToMM international conference on reconfigurable mechanisms and robots, pp 450–455

Free Vibration of Mistuned Aircraft Engine Bladed Discs

Romuald Rzadkowski and Artur Maurin

Abstract Considered here is the effect of multistage coupling on the dynamics of an aircraft engine rotor with eight mistuned bladed discs on a drum-disc shaft. Each bladed disc had a different number of rotor blades. Free vibrations were examined using finite element models of single rotating blades, bladed discs, and an entire rotor. In this study, the global rotating mode shapes of flexible mistuned bladed discs-shaft assemblies were calculated, taking into account rotational effects, such as centrifugal stiffening. The thus obtained natural frequencies of the blades, the shaft, bladed discs, and the entire shaft with discs were carefully examined and compared with a tuned system to discover resonance conditions and coupling effects. Mistuning caused considerably more intensive multistage coupling than the tuned system and distorted the mode shape nodal diameters.

Keywords Blades · Bladed discs · Free vibration · Mistuning · Multistage coupling

1 Introduction

The influence of multistage coupling on disc flexibility in mistuned bladed disc dynamics were presented for the first time by Bladh et al. (2003), who studied single-stage and two-stage rotors of a very simple stage geometry. Shahab and Thomas (1987) presented disc flexibility coupling effects on the dynamic behaviour of a multi disc-shaft system. Rzadkowski et al. (2003) showed that the coupling of three identical industrial bladed discs on a shaft segment changes the mode shapes of shrouded bladed discs up to the seventh node diameter. Sharma et al. (2005) analyzed a turbine rotor with 16 discs, with only one of them being bladed, under earthquake-force excitation, but they did not investigate couplings between the shaft

R. Rzadkowski (✉) · A. Maurin
The Szewalski Institute of Fluid-Flow Machinery, Polish Academy of Sciences, Gdansk, Poland
e-mail: z3@imp.gda.pl

A. Maurin
e-mail: amaurin@imp.gda.pl

A. Kecskeméthy and F. Geu Flores (eds.), *Interdisciplinary Applications of Kinematics*, Mechanisms and Machine Science 26, DOI 10.1007/978-3-319-10723-3_8

and bladed discs. Sinha (2007) carried out in an analysis of two mistuned bladed discs connected by a stiffness coefficient. Laxalde et al. (2007) used the multistage cyclic symmetry method to show the coupling of two bladed discs, each with a different number of blades, mounted on a flexible shaft. Rzadkowski and Drewczynski (2006, 2009, 2010) performed an analysis of eight tuned bladed discs on a shaft to show that coupling between particular bladed discs was visible up to modes with two nodal diameters. Rzadkowski and Maurin (2012) considered the influence of shaft flexibility on the dynamic characteristics of mistuned bladed discs. Free vibration analysis showed that it is important to include the shaft when investigating several mistuned bladed discs since this can considerably change the spectrum of frequencies and mode shapes with zero, one, two and more nodal diameters.

In this paper the effect of multistage coupling on the dynamics of an aircraft engine rotor with eight mistuned bladed discs on a drum-disc shaft was analyzed. Each bladed disc had a different number of rotor blades. It was found that mistuning caused considerably more intensive multistage coupling than the tuned system and distorted the mode shape nodal diameters.

2 Description of the Model

Figure 1 presents the analyzed aircraft engine rotor model. The main dimensions are as follows: the outer diameter of the largest—turbine disc is 0.512 m and the shaft length is 1.166 m. The number of blades on each disc corresponds to that of real systems. Therefore there were 28 rotor blades in compressor stage I, 41 in stage II, 41 in stage III, 47 in stage IV, 57 in stage V, 47 in stage VI, 49 in stage VII and 83 in the turbine stage.

Different mesh densities were used and a meshing convergence test based on natural frequencies was conducted. For the final calculations the smallest possible

Fig. 1 FEM model of the aircraft engine rotor

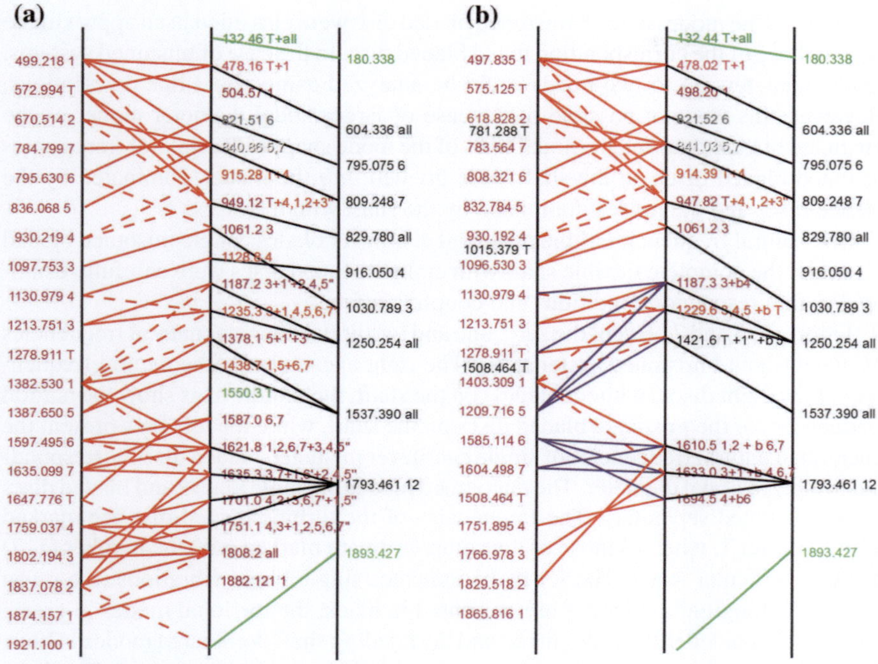

Fig. 2 Natural frequencies of tuned and mistuned bladed discs on the shaft corresponding to zero nodal diameter modes

mesh density was applied. The final FE model in its entirety had just over a 1.5 million DOFs and first natural frequency of bladed disc was 478.02 Hz (see Fig. 2). In case of 1.7 million DOFs this frequency was 478.03 Hz and in case 1.3 million DOFs was 492 Hz.

Two axis-symmetric bearing types were modeled as springs with $k_{xx} = k_{yy} = k_{xy} = k_{yx} = 1400\,\text{N}/\mu\text{m}$ stiffness for two roller bearings and $k_{xx} = k_{yy} = k_{xy} = k_{yx} = 1000\,\text{N}/\mu\text{m}$ for the ball bearing.

Two rotor models, tuned and mistuned, were developed for the numerical analyses. The blade mistuning was modeled using the modified Young modulus for every blade in each stage according to blade frequencies measured in real aircraft engine. Free vibration analyses were carried out in both models at 15,000 rpm.

3 Numerical Analysis

The natural frequencies of the rotating mistuned bladed discs were calculated using the ANSYS code. In both the tuned and mistuned models (henceforth referred to as case 1 and case 2, respectively) the natural frequencies for all the bladed discs were computed. The modes of each mistuned bladed disc were classified in an approximate way to those of the corresponding tuned bladed disc. In the case of mistuned systems, mode diameters up to two nodes could be analyzed using nodal line descriptions. However, this was not possible in the case of larger nodal diameter modes as the mistuned blades distorted the nodal lines of the mode shapes. The natural frequencies of tuned bladed discs on the shaft were divided into the modes dominated by the bladed discs and the modes dominated by the shaft with the discs.

The natural frequencies of the mistuned cantilever blades, single mistuned bladed discs and the complete flexible shaft with eight mistuned discs were carefully examined to find resonance conditions and coupling effects.

Figures 2, 5 and 7 present the zero, one and two nodal diameter natural frequencies of tuned (a) and mistuned systems (b). The right axes indicate the natural frequencies of the eight discs (without blades) on the shaft, the middle axes show the natural frequencies of the mistuned bladed discs on the shaft, while the left axes present the uncoupled natural frequencies of single cantilever mistuned bladed discs corresponding to a given nodal diameter. The uncoupled modes of single blades and bladed discs were calculated separately. The frequencies of the turbine bladed disc are marked with the letter T, whereas those of the compressor are marked with the number (1–7) of their particular stage. The letter 'b' indicates single blade vibrations in a given stage. The longitudinal modes are presented in black, the torsional modes in green, the bending bladed disc modes in red and the bending shaft dominated modes in blue. Purple lines indicate couplings between single blades and entire stages. The lines connecting the natural frequencies are divided into two types: continuous lines indicating strong coupling and dashed lines for weaker coupling. Strong coupling occurs when the amplitudes of particular blades are very visible, whereas weak coupling occurs when the amplitude is relatively small. In the latter case we observe the vibrations of a shaft and disc without blades. The frequencies and mode shapes of mistuned bladed discs on a shaft were analyzed, starting from zero nodal diameter modes.

3.1 Zero Nodal Diameters Modes

As shown in Fig. 2, the first frequency mode corresponding to the zero nodal diameter (132.46 Hz in case 1 and 132.44 Hz in case 2) was connected with a torsional rotor mode at 180.338 Hz, in both cases causing all the bladed discs to vibrate. The next frequency mode (478.16 Hz in case 1 and 478.02 Hz in case 2), presented in Fig. 3, produce coupling between the compressor first stage bladed disc and the turbine disc. These vibrations were also visible on the unbladed rotor at 604.336 Hz, thus the line connecting the two frequencies in Fig. 2. Figure 3 shows the influence of mistuning, revealing considerable differences in blade displacement between cases 1 and 2.

Similar mode couplings and distortions also occurred in subsequent zero nodal diameter modes, up to 1187 Hz (see Fig. 2). Some of the mode shapes observed in case 1 (i.e. at 1128.8 Hz etc.) did not appear in case 2 on account of the mistuning.

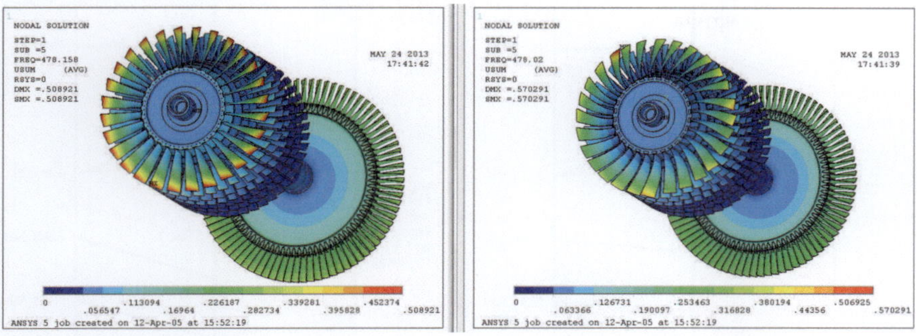

Fig. 3 Comparison of mode shapes of tuned (*left* 478.16 Hz) and mistuned (*right* 478.02 Hz) bladed discs on shaft with a zero nodal diameter

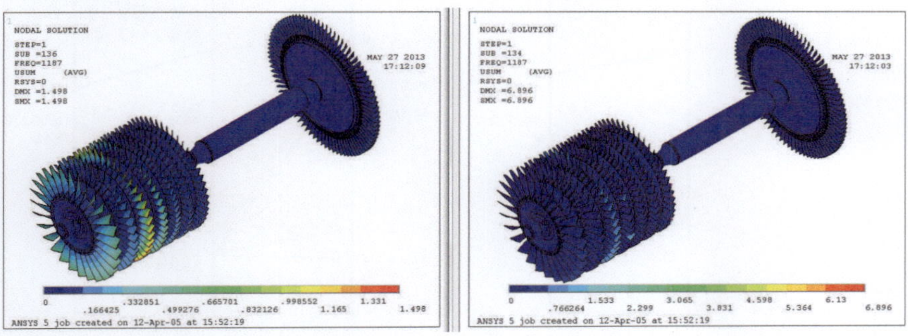

Fig. 4 Comparison of mode shapes of tuned (*left* 1187.2 Hz) and mistuned (*right* 1187.3 Hz) bladed discs on shaft with a zero nodal diameter

Generally we may conclude that mistuning alters the vibration amplitudes in particular blades. Each I stage blade in Fig. 3 has slightly different amplitude a mode with an almost zero nodal diameter. This rule,however, may not hold for other modes.When the tuned system is vibrating with a 1187.2 Hz mode coupling in the first five stages is visible (Fig. 4). Whereas in the mistuned system at 1187.3 Hz coupling is only visible in the 1st, 3rd and 4th stage with not all the blades vibrating. Moreover, the mode no longer has a zero nodal diameter.

3.2 One Nodal Diameters Modes

Figure 5 presents the natural frequencies of tuned and mistuned bladed discs (a and b) on the shaft, vibrating a one nodal diameter. The splitting of one nodal diameter double frequencies occurs in the mistuned system as an effect of blade mistuning.

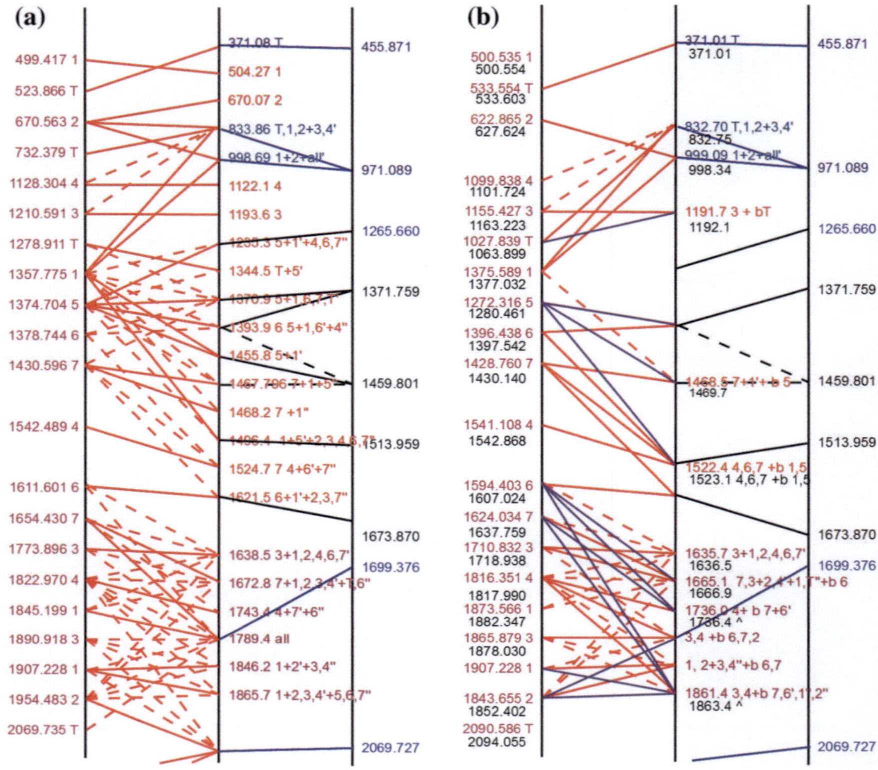

Fig. 5 Natural frequencies of tuned and mistuned bladed discs on the shaft with one nodal diameter modes

The bending shaft dominated modes are presented in blue. The 371.08 Hz mode in Fig. 5a and the 371.01 Hz mode in Fig. 5b are dominated by the bending motion of the shaft caused by the one nodal diameter mode shape of the turbine disc at 455.871 Hz. The cantilevered turbine disc in Fig. 5a vibrates at 499.417 Hz and at 533.554(533.603) Hz in Fig. 5b.

Mode couplings and distortions can be observed in subsequent one nodal diameter modes, up to the frequency of 1790 Hz (Fig. 5). Some of the mode shapes that appear in Fig. 5a (e.g. at 504.27 Hz etc.) do not occur in Fig. 5b on account of the mistuning.

Figure 6 presents 1-nodal diameter vibrations in the 3rd stage of the tuned system at 1193.6 Hz (Fig. 5) whereas in at 1191.7 Hz in the mistuned system the 1-nodal diameter mode is slightly distorted in 3rd stage and additionally individual blades vibrate in the 4th bladed disc and turbine disc.

Figure 7 present 1-nodal diameter modes in the tuned system at 1743.4 Hz and at 1736.0 (1736.4) Hz in the mistuned system (Fig. 5). In the tuned system all the compressor bladed discs are coupled, whereas in the mistuned system coupling only occurs between the 4th bladed disc and individual blades of the 6th and 7th

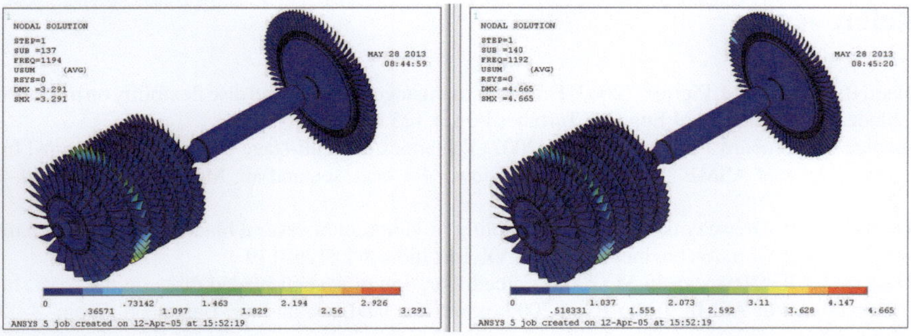

Fig. 6 Mode shapes of tuned (*left* 1193.6 Hz) and mistuned (*right* 1191.7 Hz) bladed discs on shaft with one nodal diameter

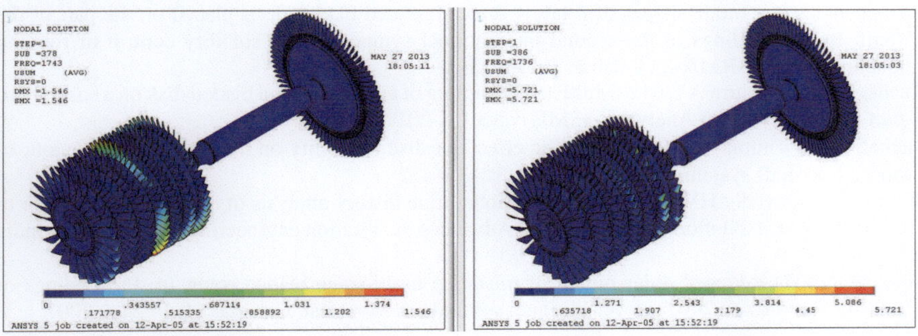

Fig. 7 Mode shapes of tuned (*left* 1743.4 Hz) and mistuned (*right* 1736.0 Hz) bladed discs on shaft with one nodal diameter

bladed discs. A similar situation occurs in mistuned modes at 1468.5 (1469.7), 1522.4 (1523.1), 1665.1(1666.9), and 1861.4 (1863.4) Hz.

We may conclude that the higher the mode, the greater the influence of mistuning.

4 Conclusions

Free vibration analysis of an aircraft rotor with 8 mistuned bladed discs showed that mistuning considerably increases multistage coupling when compared with a tuned rotor system.The mistuning distorts the nodal diameters of mode shapes and causes multistage coupling between individual blades.

Free vibration analysis has also shown that mistuning changes the mode shapes.

Due to mistuning, the number of resonances and couplings in the system may change. Free vibration analysis has not sufficiently explained multistage coupling in tuned bladed discs. To better understand this problem the forced vibration analyses should be carried out.

References

Bladh R, Castanier M, Pierre C (2003) Effects of multistage coupling and disc flexibility on mistuned bladed disk dynamics. J Eng Gas Turbines Power 125:121–130

Laxalde D, Lombard J-P, Thouverez F (2007) Dynamics of multi-stage bladed disks systems. In: Proceedings of ASME Turbo Expo 2007: power for land, sea and air, Montreal, Canada, 14–1 May (2007)

Rzadkowski R, Drewczynski M (2009) Coupling of vibration of several bladed discs on the shaft. Advances in vibration engineering. Sci J Vib Inst India 8(2):125–139

Rzadkowski R, Drewczynski M (2006) Forced vibration of several bladed discs on the shaft. In: Proceedings of ASME Turbo Expo 2006: power for land, sea and air, Barcelona, Spain, 8–11 May 2006

Rzadkowski R, Drewczynski M (2010) Multistage coupling of eight bladed discs on a solid shaft. In: Proceedings of ASME Turbo Expo 2010: power for land, sea and air, Glasgow, UK, 14–18 June (2010)

Rzadkowski R, Kwapisz L, Sokołowski J, Karpiuk R, Ostrowski P, Radulski W (2003) Natural frequencies and mode shapes of rotating three shrouded blades discs placed on the part of the shaft. In: Proceedings of the second international symposium on stability control of rotating machinery (ISCORMA-2), Gdansk, pp 381–392

Rzadkowski R, Maurin A (2012) Multistage coupling of eight mistuned bladed disk on a solid shaft, part 1. Free Vibration Analysis. ASME paper GT-2012-68391

Shahab AAS, Thomas J (1987) Coupling effects of disc flexibility on the dynamics behaviour of multi disc-shaft system. J Sound Vib 114(3):435–452

Sharma BK, Devadig HV, Singh K (2005) Modal time history analysis of a steam turbine rotor to an earthquake excitation: a 3D approach. Advances in vibration engineering. Sci J Vib Inst India 4(4):351–359

Sinha A (2007) Reduced-order model of mistuned multi-stage bladed rotor. In: Proceedings of ASME Turbo Expo 2007: power for land, sea and air, Montreal, Canada, 14–17 May 2007

On the Study of the Kinematic Position Errors Due to Manufacturing and Assembly Tolerances

Paulo Flores

Abstract This work deals with the kinematic position errors analysis of planar multibody mechanical systems when considering tolerances associated with manufacturing and assemble processes. For this purpose, the generalized Cartesian coordinates are utilized to mathematically formulate kinematic constraints and equations of motion of the multibody systems. The systems are defined by a set of generalized coordinates, which represents the instantaneous positions of all bodies, together with a set of generalized dimensional parameters that defines the functional dimensions of the system under analysis. The generalized dimensional parameters can take into account the tolerances associated with the lengths, fixed angles, diameters and distance between centers. This work emphasizes the relation among kinematic constraints, dimensional parameters and output kinematic parameters. Based on the theory of dimensional tolerances, the variation of the geometrical dimensions is regarded as a tolerance grade with an interval associated with each dimension and, consequently, a kinematic amplitude variation for the position of bodies.

1 Introduction

In general, there are two main approaches to study the effect of the manufacturing tolerances on the kinematic position errors, namely, the deterministic and probabilistic methods. The deterministic method involves fixed values or constraints that are used to find an exact solution. These methods are used mostly when tolerances are known and the worst position error is to be determined. In contrast, the probabilistic or statistical methods deal with random variables that result in a probabilistic response. The statistical approaches are utilized when dimensions have some type of random distribution and the probability of being within a given tolerance band is to be evaluated.

Tolerances play a key role in the modern design process by introducing quality improvements and limiting manufacturing costs. According to standard ANSI

P. Flores (✉)
Mechanical Engineering Department, University of Minho, Braga, Portugal
e-mail: pflores@dem.uminho.pt

A. Kecskeméthy and F. Geu Flores (eds.), *Interdisciplinary Applications of Kinematics*, Mechanisms and Machine Science 26, DOI 10.1007/978-3-319-10723-3_9

Y14.5M-1994, tolerances are used to define the allowable limits of geometric variation that are inherent in manufacturing and assemble processes ANSY (1994). The assignment of geometric tolerances is always a trade-off between two distinct situations: (i) a part with tight tolerances is good for assembly, but the cost to manufacture the part is increased; (ii) alternatively, loose tolerance in one part may make the whole assemble infeasible. The tolerance analysis deals with the study of the aggregate behaviors of given individual tolerances (Lee and Woo 1990; Di Stefano 2006).

Over the last few decades a number of research papers has been published on the influence of the tolerance and clearance on the kinematic performance of multibody systems (Garrett and Hall 1969; Dhande and Chakraborty 1973; Hummel and Chassapis 2000; Shi 1997; Choi et al. 1998; Wittwer et al. 2004; Fogarasy and Smith 1998; Chase and Parkinson 1991). However, most of these research works lack of generality, that is, they are developed for specific mechanisms and situations. Garrett and Hall (1969) defined the concept of mobility bands to study the effect of tolerance and clearance in the design of linkages. Dhande and Chakraborty (1973) presented a stochastic model for the analysis and synthesis of the four-bar mechanism considering tolerances and clearances. Hummel and Chassapis (2000) described an approach to the design and optimization of Cardan joints with manufacturing tolerances. Based on the reliability concept, Shi (1997) presented and developed a probabilistic model of mechanical error in spatial mechanisms. Choi et al. (1998) presented an analytical approach to tolerance optimization for planar mechanisms with lubricated joints based on mechanical error analysis. Wittwer et al. (2004) established the direct linearization method applied to position error in kinematic linkages due to the link-length and angle variation. Fogarasy and Smith (1998) presented a complete investigation on the study of the influence of manufacturing tolerances on the kinematic response of mechanisms. A good survey on the research work developed in the field of tolerance analysis of kinematic mechanism is provided by Chase and Parkinson (1991).

It is well known that, the manufacturing cost increases for uniform incremental tightening of tolerances ISO (1988). The cost is also related to the characteristics of manufacturing processes utilized, and the degree of maturity of the workers. Tolerances should be allocated as large as possible for the sake of economy and ease of manufacture. However, large tolerances usually increase mechanical errors. Thus, designers should allocate tolerances to minimize manufacturing cost while keeping mechanical errors below a certain specific limit. The purpose of this work is to present a general approach to quantify the kinematic position errors due to manufacturing and assemble tolerances. The kinematic constraints and equations of motion of the multibody systems are formulated under the framework of multibody systems formalism. The system is defined by set of generalized coordinates, which represents the instantaneous positions of all bodies, together with a set of generalized dimensional parameters that defines the functional dimensions of the system. The generalized dimensional parameters take into account the tolerances associated with the lengths. The relation between the kinematic constraints, dimensional parameters and output kinematic parameters is demonstrated.

2 Kinematic Analysis

When the configuration of a multibody system is described by n Cartesian coordinates, then a set of algebraic kinematic independent holonomic constraints $\mathbf{\Phi}$ can be written in a compact form as

$$\mathbf{\Phi}(\mathbf{q}, t) = \mathbf{0} \tag{1}$$

where $\mathbf{q}$ is the vector of generalized coordinates and t is the time variable, usually associated with the driving elements.

The velocities and accelerations of the system elements are evaluated through the velocity and acceleration constraint equations. Thus, the first time derivative with respect to time of Eq. (1) provides the velocity constraint equations

$$\mathbf{\Phi}_{\mathbf{q}}\dot{\mathbf{q}} = -\mathbf{\Phi}_t \equiv \boldsymbol{\upsilon} \tag{2}$$

where $\mathbf{\Phi}_{\mathbf{q}}$ is the Jacobian matrix of the constraint equations, $\dot{\mathbf{q}}$ is the vector of generalized velocities and $\boldsymbol{\upsilon}$ is the right hand side of velocity equations.

A second differentiation of Eq. (1) with respect to time leads to the acceleration constraint equations, obtained as

$$\mathbf{\Phi}_{\mathbf{q}}\ddot{\mathbf{q}} = -(\mathbf{\Phi}_{\mathbf{q}}\dot{\mathbf{q}})_{\mathbf{q}}\dot{\mathbf{q}} - 2\mathbf{\Phi}_{\mathbf{q}t}\dot{\mathbf{q}} - \mathbf{\Phi}_{tt} \equiv \boldsymbol{\gamma} \tag{3}$$

where $\ddot{\mathbf{q}}$ is the acceleration vector and $\boldsymbol{\gamma}$ is the right hand side of acceleration equations, i.e., the vector of quadratic velocity terms. In the case of holonomics cleronomic constraints, that is, when $\mathbf{\Phi}$ is not explicitly dependent on the time, the term $\mathbf{\Phi}_t$ in Eq. (2) and the $\mathbf{\Phi}_{\mathbf{q}t}$ and $\mathbf{\Phi}_{tt}$ terms in Eq. (3) vanish.

The constraint equations represented by Eq. (1) are, in general, nonlinear in terms of $\mathbf{q}$ and are usually solved by the Newton-Raphson method. Equations (2) and (3) are linear in terms of $\dot{\mathbf{q}}$ and $\ddot{\mathbf{q}}$, respectively, and can be solved by any usual method for linear equations' systems. Thus, the kinematic analysis of a multibody system can be carried out by solving the set of Eqs. (1), (2) and (3).

3 Kinematic Position Errors

In order to evaluate, in a systemic and general way, the influence of the manufacturing and assemble tolerances on the kinematic position errors, special attention needs to be given to the mathematical formulation of the description of the systems' configuration. Thus, according to the previous section, the equations of constraints can be written as

$$\mathbf{\Phi}(\mathbf{q}_1, \mathbf{q}_2, \ldots, \mathbf{q}_n, \mathbf{d}_1, \mathbf{d}_2, \ldots, \mathbf{d}_m) = \mathbf{0} \tag{4}$$

where $\mathbf{q}_1, \mathbf{q}_2, \ldots, \mathbf{q}_n$ represent the generalized vector of coordinates that define the kinematic system's configuration at any instant, and $\mathbf{d}_1, \mathbf{d}_2, \ldots, \mathbf{d}_m$ are the generalized vectors of the dimensional parameters defining the functional dimensions of the system. It should be noted that Eq. (4) represents the kinematic system's constraints, which can easily be written using, for instance, Cartesian coordinates. Furthermore, the number of generalized coordinates, n, and the number of generalized dimensional parameters, m, must be adequately selected bearing in mind the correct system's description and system's degrees of freedom.

The kinematic analysis of any multibody system implies the resolution of Eq. (4) for $\mathbf{q}_1, \mathbf{q}_2, \ldots, \mathbf{q}_n$ and their derivatives, according to what was presented in the previous section. In this process, it is assumed that vectors $\mathbf{d}_1, \mathbf{d}_2, \ldots, \mathbf{d}_m$ are constants, meaning that there is no variation of the dimensional parameters and, consequently, they do not affect the global system's performance. However, it is well known that it is not the case in practical engineering design and manufacturing processes. Since, in general, the multibody systems are conducted by driving elements, excluding these elements and considering that the kinematic constraints are independent, the Jacobian matrix can be written as follows

$$\mathbf{\Phi q} = \frac{\partial \mathbf{\Phi}_k}{\partial \mathbf{q}_l}, \quad (k = 1, \ldots, n - dr;\ l = 1, \ldots, n) \tag{5}$$

where indices k and l are, respectively, the $n - dr$ kinematic constraints and n Cartesian coordinates. The number of driving elements is represented by variable dr.

Considering all coordinates and dimensional parameters as global system's variables, the variation of the constraint equation is expressed as

$$\frac{\partial \mathbf{\Phi}_k}{\partial \mathbf{q}_1}\delta\mathbf{q}_1 + \cdots + \frac{\partial \mathbf{\Phi}_k}{\partial \mathbf{q}_n}\delta\mathbf{q}_n + \frac{\partial \mathbf{\Phi}_k}{\partial \mathbf{d}_1}\delta\mathbf{d}_1 + \cdots + \frac{\partial \mathbf{\Phi}_k}{\partial \mathbf{d}_m}\delta\mathbf{d}_m = \mathbf{0} \tag{6}$$

In a compact form, Eq. (6) can be written as

$$\sum_{i=1}^{n-dr} \frac{\partial \mathbf{\Phi}_k}{\partial \mathbf{q}_i}\delta\mathbf{q}_i + \sum_{j=1}^{m} \frac{\partial \mathbf{\Phi}_k}{\partial \mathbf{d}_j}\delta\mathbf{d}_j = \mathbf{0}, (k = 1, \ldots, n - dr) \tag{7}$$

in which, $\delta\mathbf{q}_i$ is the variation of the generalized system's coordinates and $\delta\mathbf{d}_j$ is the variation of the dimensional parameters. This last term represents the manufacturing tolerances of the corresponding functional dimensions, such as the lengths of the multibody system's parts. In a matrix form, Eq. (7) is expressed as

$$\mathbf{\Phi}_{\mathbf{q}_i}\delta\mathbf{q}_i = -\mathbf{\Phi}_{\mathbf{d}_j}\delta\mathbf{d}_j, \quad (i = 1, \ldots, n - dr;\ j = 1, \ldots, m) \tag{8}$$

where $\mathbf{\Phi}_{\mathbf{q}_i}$ is the Jacobian matrix and $\mathbf{\Phi}_{\mathbf{d}_j}$ represents the derivative of the constraint equations in relation to the dimensional parameters. Since in kinematic analysis of multibody systems the Jacobian matrix is known, as illustrated in the previous section, specifying the manufacturing tolerances, $\delta\mathbf{d}_j$, only the matrix $\mathbf{\Phi}_{\mathbf{d}_j}$ needs to be evaluated in order to obtain the kinematic position errors of all system's bodies, by solving Eq. (8). It should be noted, that with this approach very low computational effort is added to the standard kinematic analysis procedure. Moreover, it should be highlighted that Eq. (8) represents a linear system of equations that can easily be solved by employing any numerical method, such as the LU factorization procedure.

In order to quantify the kinematic accurate position of a system part, it is first necessary to define the amount of tolerance allowed for each one of the dimensional parameters considered. Thus, according to standard ISO 286-1, for common mechanical operating conditions the IT grades are usually in the range IT8 to IT11 ISO (1988). For the manufacturing tolerances of the dimensional parameters of a multibody system, the bilateral tolerances specified in ISO 286-1 are commonly used. Hence, in Eq. (8), the variation of the functional parameters $\delta\mathbf{d}$can be regarded as such tolerance fields. Therefore, it is possible to write the following relation

$$\delta\mathbf{d} = \pm\frac{1}{2}T \tag{9}$$

where T represents the total manufacturing tolerance range corresponding to the dimensional parameters.

4 Manufacturing and Assemble Tolerance Analysis

It is known that tolerances are used to define the allowable limits of geometric variations that are inherent in the manufacturing and assembly processes ANSY (1994). Broadly, there are two approaches to solve this problem; worst case assemblability and statistical assemblability. In the first case, it is assumed that all the component dimensions occur, in each assembly, at their extreme and worst limit simultaneously. When this approach is employed, the designer desires to ensure that the components can always be assembled. This means that the probability of having a kinematic error exceeding the specific limits in a particular system is null. On the other hand, the statistical assemblability can be used to take advantage of statistical averaging over of components, allowing for the use of less restrictive tolerances in exchange for admitting the small probability of non-assembly. In general, the standard process is defined at the confidence level corresponding to $\pm 3\ \sigma$ interval (Ang and Tang 1984a, b). From the statistical point of view the worst case model is the most pessimistic, while the statistical assemblability is the most optimistic case. In practical situations it is expected that they fall between the worst and statistical models. These two approaches are discussed in detail in the following paragraphs. Using Eq. (8) as reference, the variation of the generalized system's coordinates can be rewritten as

$$\delta\mathbf{q} = \mathbf{s}\delta\mathbf{d} \tag{10}$$

where **s** represents the sensitive coefficients given by

$$\mathbf{s} = -\mathbf{\Phi}_{\mathbf{q}}^{-1}\mathbf{\Phi}_{\mathbf{d}} \tag{11}$$

Considering a system with *m* generalized dimensional parameters and that the maximum tolerance or error is specified, based on the worst scenario assemblability, it is possible to evaluate the maximum error of a general output coordinate as

$$\frac{1}{2}\mathrm{T_{maximum}} \geq \delta\mathbf{q} = \sum_{j=1}^{m} \left|s_j\right| \frac{1}{2}\mathrm{T_j} = \frac{1}{2}\left(s_1\mathrm{T_1} + s_2\mathrm{T_2} + \cdots + s_m\mathrm{T_m}\right) \tag{12}$$

The average tolerance can be determined by the following expression

$$\mathrm{T_{average}} = \frac{\sum_{j=1}^{m} |s_j| \frac{1}{2}\mathrm{T_j}}{m} = \frac{\mathrm{T_{max}}}{m} \tag{13}$$

In a similar way, the tolerance associated with any system component can be given by

$$\mathrm{T_i} = \frac{\mathrm{T_{max}}}{s_i m} \tag{14}$$

It is worth noting that $\mathrm{T_{maximum}}$ is a given quantity, allowing the evaluation of average tolerance values, which ultimately can be used as guiding reference to specify the manufacturing precision requirements of the system components by selecting the IT grades.

Tolerance is statistical in nature since the output of a random variable is, in general, normally or Gaussian distributed, with the level of confidence three-sigma considered. This means that only two or three cases in a thousand have the probability to be outside the $\pm 3\sigma$ range. This confidence level of tolerance becomes an important design parameter to be evaluated and optimized. Thus, using the statistical approach, the root mean square considers that the component dimensions occur statistically having a Gaussian distribution and can be expressed by

$$\frac{1}{2}\mathrm{T_{maximum}} \geq \delta\mathbf{q} = \sqrt{\sum_{j=1}^{m} \left|s_j\right|^2 \left(\frac{1}{2}\mathrm{T_j}\right)^2} \tag{15}$$

In the statistical approach, the average tolerance can be determined by the following expression

$$\mathrm{T_{average}} = \frac{\mathrm{T_{max}}}{\sqrt{m}} \tag{16}$$

Table 1 Values of γ factor for different confidence levels

$p = 99.7\,\%$	$f(\gamma) = 0.997$	$\gamma = 2.96$
$p = 99.0\,\%$	$f(\gamma) = 0.990$	$\gamma = 2.58$
$p = 97.0\,\%$	$f(\gamma) = 0.970$	$\gamma = 2.17$
$p = 95.0\,\%$	$f(\gamma) = 0.950$	$\gamma = 1.96$

In a similar way, the tolerance associated with any system component can be given by

$$T_i = \frac{T_{max}}{s_i\sqrt{m}} \tag{17}$$

In order to convert the tolerance value to the standard process-tolerance, the γ factor is introduced Di Stefano (2006)

$$\gamma = \frac{T_{maximum}}{2\sigma} \tag{18}$$

Solving Eq. (18) for σ yields

$$\sigma = \frac{T_{maximum}}{2\gamma} \tag{19}$$

Consequently, the admissible tolerance of any component can be expressed by

$$T_{admissible} = 6\sigma = \frac{3T_{maximum}}{\gamma} \tag{20}$$

In Table 1, the γ factor is listed as a function of confidence p value of the tolerance. This table was constructed by integrating the standard normal distribution function $f(\gamma)$, which can be defined by the error integral in the form Ang and Tang (1984a, b)

$$\frac{p}{100} = f\,(\gamma) = \frac{1}{2\pi}\int_{-\gamma}^{\gamma} e^{t^2/2} d\,\gamma \tag{21}$$

5 Demonstrative Example of Application

An elementary slider-crank mechanism is used here to show the influence of manufacturing tolerances on kinematic performance. Figure 1 shows the configuration of the mechanism, comprising four rigid bodies that represent the crank, connecting rod, slider and ground, three revolute joints and one ideal translational joint. The inertia properties and length characteristics of each body, as well as the associated tolerance according to ISO 286-1 standard are given in Table 2. In the present example the

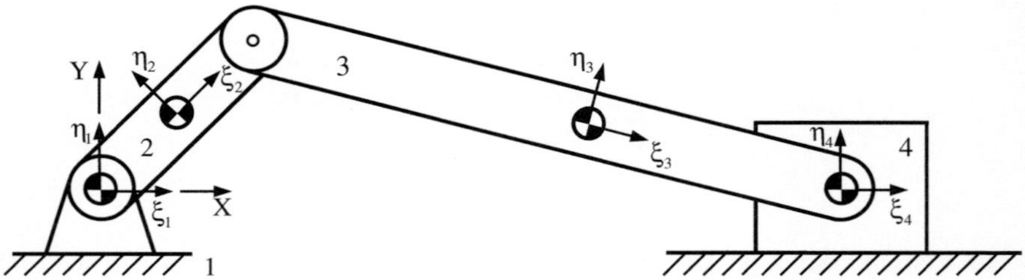

Fig. 1 Schematic representation of the slider-crank mechanism

Table 2 Geometric properties of the slider-crank mechanism

Body nr	Description	Length (m)	Tolerance range (μm)
1	Ground	–	–
2	Crank	0.050	± 50
3	Connecting rod	0.120	± 70
4	Slider	–	–

tolerance grade IT 10 was selected establishing the number of generalized dimensional parameters equal to two ($m = 2$), which are related to the lengths of the crank (r_2) and connecting rod (r_3).

In the kinematic simulation, the crank is the driving element and rotates at a constant angular velocity of 500 rpm clockwise. The initial system configuration corresponds to the top dead point. In the numerical simulation, r_2 and $r3$ were selected as dimensional tolerance parameters, being δx_4 and $\delta\theta_3$ the output parameters. Thus, applying the methodologies presented in the previous sections, Figs. 2 and 3 show, respectively, the maximum absolute errors on the linear slider position and the angular position of the connecting rod, when the worst case approach is considered. These

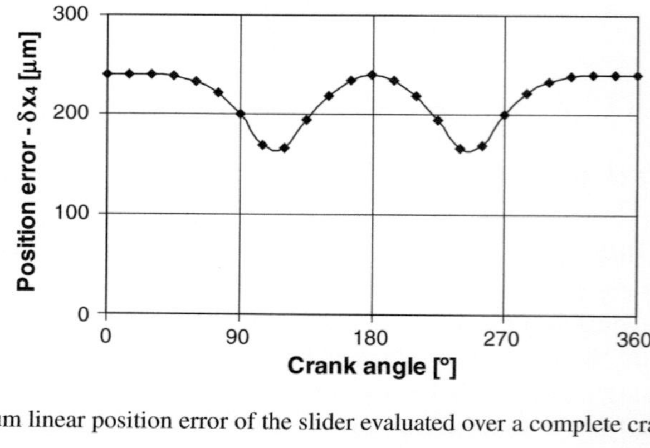

Fig. 2 Maximum linear position error of the slider evaluated over a complete crank cycle

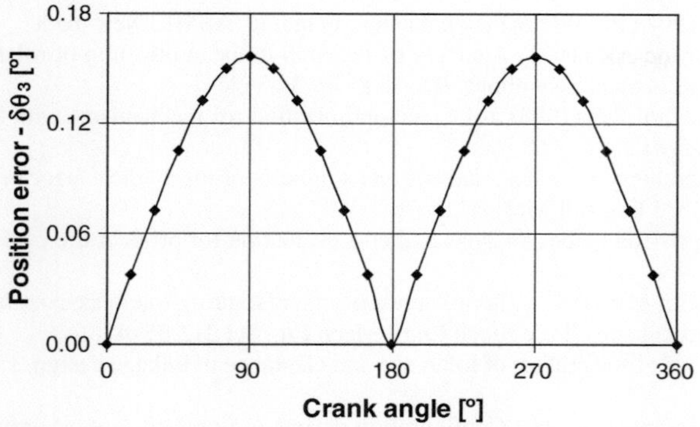

Fig. 3 Maximum angular position error of the connecting rod evaluated over a complete crank cycle

maximum position errors were evaluated at 25 crank angular positions. By observing the global results obtained it can be concluded that the maximum position errors vary during the computational simulation of the slider crank mechanism.

6 Conclusions

In this paper, a general and systematic methodology for kinematic positional error analysis of multibody systems was investigated, taking into account the influence of the manufacturing and assembly tolerances on the performance of planar mechanisms. In the process, the main aspects for kinematic analysis of multibody systems were revised. Based on the theory of dimensional tolerances, the variation of the geometrical dimensions is regarded as a tolerance grade with an interval associated with each dimension and, consequently, a kinematic amplitude variation for the positions. The presented deterministic method evaluates the relation between variations in the dimensional parameters and variation in the generalized coordinates. The statistical approach based on the confidence level three-sigma was also studied. The methodologies proposed have been exemplified through the application of kinematicsto a slider-crank mechanism. The simplicity and generality of the proposed methodology for the study of kinematic position errors due to manufacturing tolerances was thus demonstrated.

References

Ang AHS, Tang WH (1984a) Probability concepts in engineering planning and design. Vol. I: Basic Principles. John Wiley and Sons, New York

Ang AHS, Tang WH (1984b) Probability concepts in engineering planning and design. Vol. II: Decision, Risk, and Reliability. John Wiley and Sons, New York

ANSY, Y14.5M-1994 (1994) Dimensional and tolerancing. ASME, New York

Chase KW, Parkinson AR (1991) A survey of research in the application of tolerance analysis to the design of mechanical assemblies. Res Eng Des 3:23–37

Choi JH, Lee SJ, Choi DH (1998) Tolerance optimization for mechanisms with lubricated joints. Multibody Sys Dyn 2:145–168

Dhande SG, Chakraborty J (1973) Analysis and synthesis of mechanical error in linkages: A stochastic approach. J Eng Ind 95(3):672–676

Di Stefano P (2006) Tolerances analysis and cost evaluation for product life cycle. Int J Prod Res 44:1943–1961

Fogarasy AA, Smith MR (1998) The influence of manufacturing tolerances on the kinematic performance of mechanisms. Proc Mech Eng J Mech Eng Sci 212:35–47

Garrett RE, Hall AS (1969) Effect of tolerance and clearance in linkage design. J Eng Ind 91:198–202

Hummel SR, Chassapis C (2000) Configuration design and optimization of universal joints with manufacturing tolerances. Mech Mach Theory 35:463–476

ISO 286-1 (1988) ISO system of limits and fits—Parts 1: Base of tolerances, deviations and fits

Lee WJ, Woo TC (1990) Tolerances: their analysis and synthesis. J Eng Ind 112:113–121

Nikravesh PE (1988) Computer aided analysis of mechanical systems. Prentice Hall, Englewood Cliffs

Shi Z (1997) Synthesis of mechanical error in spatial linkages based on reliability concept. Mech Mach Theory 32(2):255–259

Wittwer JW, Chase KW, Howell LL (2004) The direct linearization method applied to position error in kinematic linkages. Mech Mach Theory 39:681–693

Kinematic Design Problems for Low-Cost Easy-Operation Humanoid Robots

Marco Ceccarelli

Abstract In this paper design problems are discussed for developing low-cost easy-operation humanoid robots by looking at a kinematic solution with a reduced number of d.o.f.s and mechanism structure for fairly simple operation. The outlined design problems have been attached to propose a new kinematic solution that is under development at LARM in Cassino.

Keywords Humanoid robots · Kinematic design · Parallel manipulators

1 Introduction

Humanoid robots have been designed and built in the last decades as much as the technology has permitted complex integration of multi-body systems with suitable capabilities in actuators, sensor, control equipment, artificial intelligence, and interfaces, (Android World homepage 2013). Very advanced but sophisticates prototypes have been built, like for example Honda Asimo (Honda Motor Co 2013), Wabian (HRI Takanishi lab 2013), HRP (KAWADA Industries Inc 2013). They have been used within lab environments, but even in some applications, with the main purposes of exploring design solutions and performance capabilities in mimicking humans and their actions. Recently, solutions have reached market availability with user-oriented performance, like for example Nao (Robotics 2013), Pino (JST 2013), although they still require high expertise for a proper operation and use.

In general the activities for humanoid robots can be understood as focused along two main directions, namely design developments of sophisticated solutions towards fully android behaviors, and design arrangements of limited-skill solutions towards low-cost easy-operation systems. Most interesting advances are worked out in project developments for sophisticated solutions, like for example for the above mentioned Honda Asimo, Wabian, HRP. But those advances are also used in the second category of approaches where challenges are attached to achieve humanoid designs and

M. Ceccarelli (✉)
University of Cassino and South Latium, Via Di Biasio 43, 03043 Cassino, Italy
e-mail: ceccarelli@unicas.it

A. Kecskeméthy and F. Geu Flores (eds.), *Interdisciplinary Applications of Kinematics*, Mechanisms and Machine Science 26, DOI 10.1007/978-3-319-10723-3_10

operations with less complicated solutions towards market solutions even for specific applications, like guiding and surveillance.

This paper refers to an activity at LARM in Cassino that is aimed to develop a low-cost user-oriented humanoid robot with limited skills as based on mechanism design with fairly simple controlled operation. In the paper problems and requirements are discussed with main aspects that can motivate and justify the LARM solution that is proposed herein with a kinematic design.

2 Design Requirements and Operation Features

In general, humanoid robots are developed to fully mimic the structure and operation capabilities of human beings within research and development projects both with application purposes and study aims.

Main requirements and design structures for a humanoid robot can be summarized as in Fig. 1, in which a segmentation of the system emphasizes primary functions of the subsystems, as specifically focused on mechanical properties related to human performance. A numerical estimation of the merits of those performances is summarized in Table 1 as related to d.o.f. structure for driving power units and in Table 2 as related to operation capability in main tasks. Data of human anatomy are summarized as related to main aspects that are used in designing and operating humanoid robots.

Expected data for humanoid robots are related to a desired mimic of human design with human-like operation capability, as developed in the most advanced built prototypes.

Data for reduced solutions are related to solutions that are aimed at specific humanoid applications with a considerable reduction of design sophistication and operation complexity.

It is understood that a replica of human body is still far to be possible in terms of efficiency and flexibility of motion capabilities with sensored controlled compactness. It is also well understood that in many applications, mainly with today solutions, a

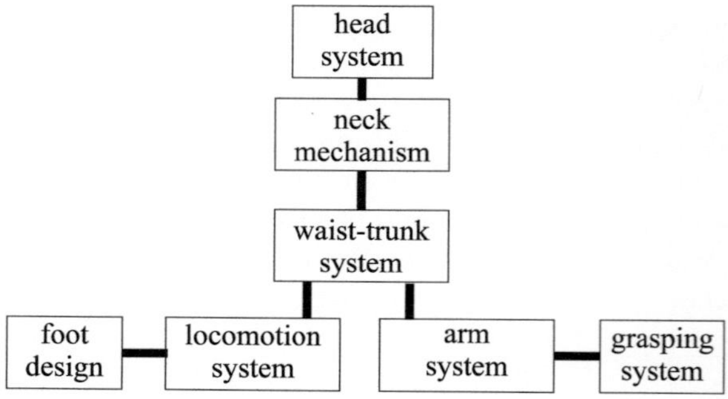

Fig. 1 A general scheme of sub-system composition of a humanoid robot

Table 1 Main d.o.f.s for humanoid robot designs

Sub-system	Human anatomy	Expected humanoid	Reduced solution
Foot	4	3	2
Leg	3	3	2
Waist	3	2	1
Trunk	3	3	2
Arm	4	3	2
Wrist	3	3	2
Hand	21	10	1
Neck	3	3	2
Head	5	4	2

Table 2 Main operation features for humanoid robots

Operation task	Humanoid sub-system	Human anatomy	Expected humanoid	Reduced solution
Ground contact	Foot	4	3	2
Walking	Leg	3	3	2
Walking stability	Waist	3	2	1
Payload capability	Trunk	3	3	2
Manipulation	Arm	4	3	2
Orientation	Wrist	3	3	2
Grasping	Hand	21	10	1
Head orientation	Neck	3	3	2
Sensoring and expression	Head	5	4	2

humanoid robot will not reproduce all the human-like functionalities. Thus, although at research level a full humanoid robot is still a challenging goal, humanoid robots with reduced skills can be developed and applied successfully for specific tasks of human-like behavior. Requirements can be identified in the characteristics of specific tasks with the aim to identified structure and equipment for the application-oriented design. Similarly, operation features can be identified for a user-oriented operation with suitable limited performance.

An interesting reduced solution can be addressed for tasks with basic capabilities in walking and carrying limited loads for applications like surveillance/inspection, guiding (in museum and entertainment environments, or assistance to disabled people), house service, and entertainments.

This paper is focused on a reduced solution of humanoid robot with a mechanism-based structure for a limited number of servo controlled actuators.

Humanoid robots are based on mechanical designs as function of the planned biomechanical tasks. But more and more humanoid behavior and performance are achieved in advanced solutions with strong contribution from components of other engineering disciplines, like control, vision, measuring, computer science and so on. Nevertheless a mechanical structure is the core of humanoid robots. Mechanism-based solutions can be thought by using mechanisms that give functionality and payload capability with human-like behaviors. Usually the mechanisms that are used for the structure of humanoid robots are fairly simple ones with the aim to produce a compact serial kinematic chain.

In this paper alternative solutions are investigated to propose kinematic designs whose functionality is given by mechanism solutions rather than actuation laws, without limiting the compactness of the mechanical design when with kinematic serial chains.

3 A New Kinematic Design

Mechanism can be used in humanoid structure components for their mechanical efficiency, payload capability, and input-output features. In addition, a use of well known mechanisms can facilitate understanding but acceptance of humanoid robots and their use, even when they have not a human-like appearance.

Mechanism solutions have been proposed as in the following for each humanoid sub-structured by looking at previous experiences and design at LARM in Cassino.

A bipedal locomotion system can be designed by using two different approaches for leg design, namely linkage-based mechanisms and parallel manipulator architectures. In Fig. 2 attempted solutions at LARM are shown as by inspiration from the human structure and functionality, Fig. 2a. In Fig. 2b a prototype is shown with a linkage design combining a pantograph with a Chebyshev 4-bar for a suitable foot point trajectory and its amplification, (Li and Ceccarelli 2011). In Fig. 2c a design is proposed with parallel manipulator architecture with a powerful structure for heel function and light independent trusses for foot motion regulation, (Ceccarelli and Carbone 2009).

The wais-trunk system can be modeled through main reference platforms as in Fig. 3a so that convenient solutions have been experienced successfully by using 3 d.o.f. CAPAMAN design as in Fig. 3b, (Nava Rodriguez et al. 2006), and by developing a powerful human-like solution as in Fig. 3c, with a 3 d.o.f. waist and 6 d.o.f. trunk, (Liang and Ceccarelli 2012).

In Fig. 4a a structure of arm is shown with a human-like 3 d.o.f. design with actuators in the shoulder joint and in Fig. 4b the kinematic design of LARM clutched arm, (Gu et al. 2010) is reported with a scheme of 1 d.o.f. actuation via clutches.

In Fig. 5a the bone structure of human hand is illustrated as reference for the kinematic design of fingers in the LARM Hand, Fig. 5b, (Carbone and Ceccarelli

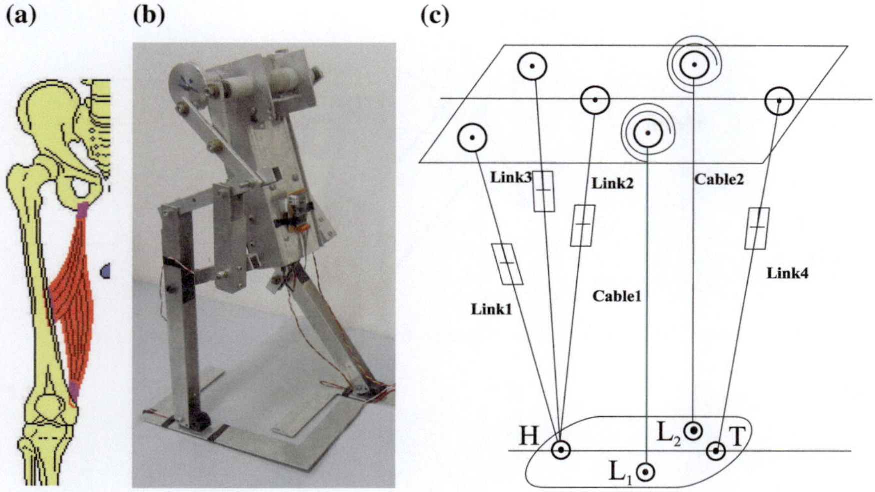

Fig. 2 Model for the locomotion system: **a** human structure, **b** LARM linkage-based biped, **c** a leg design with parallel manipulator architecture

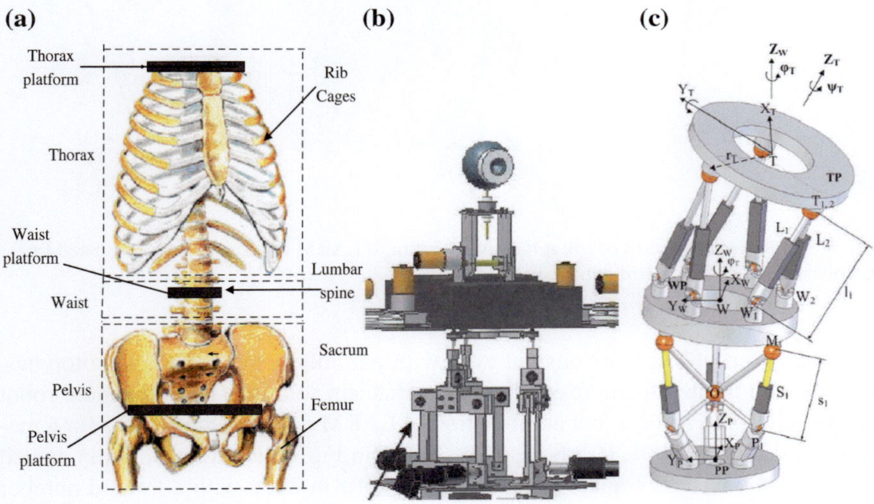

Fig. 3 Model for the waist-trunk system: **a** human structure, **b** CAPAMAN-based LARM trunk, **c** a waist-trunk system with parallel manipulator architecture

2008), in which each finger is driven by a cross-4 bar linkage with one actuator. Figure 5c shows a kinematic design of an underactuated mechanism, (Wu et al. 2009), to improve the finger adaptability in object grasping.

The above-mentioned sub-systems have been developed and experienced as independent design, with prototypes at LARM. Studies of feasibility have been worked

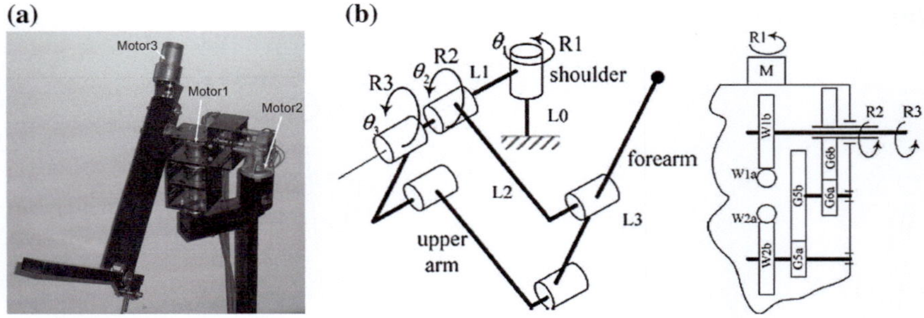

Fig. 4 Model for the 3 d.o.f. arm system: **a** human-like design, **b** LARM clutched arm

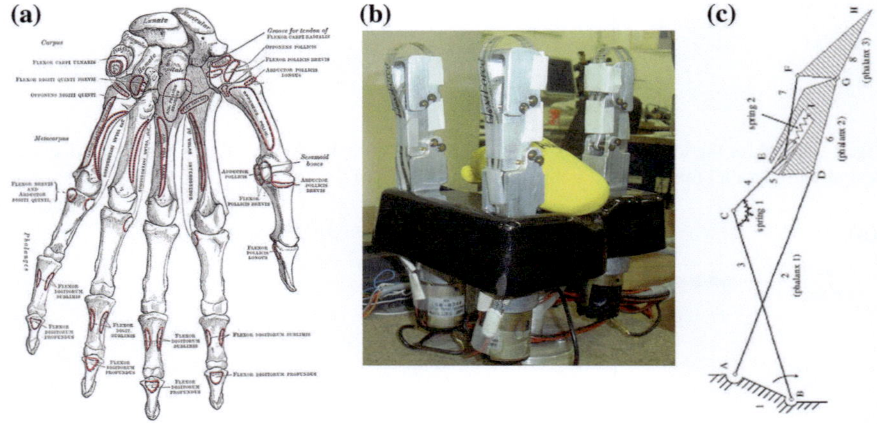

Fig. 5 Model for the hand system: **a** human structure, **b** LARM hand, **c** an undertactuated finger mechanism for a new LARM hand

out mainly through simulations but even with partially combination of prototypes, with the aim to investigate feasibility of mechanism solutions in a humanoid robot design. Thus, in Fig. 6a a first combination of LARM designs is presented for a mechanical design in 2001, (Carbone et al. 2001); in Fig. 6b a refined solution named as CALUMA is shown again with LARM designs in 2006 with a limited number of actuated d.o.f.s for an on-board PLC control, (Nava Rodriguez et al. 2006); in Fig. 6c a mechanical design is proposed in 2011 as centered on waist-trunk system with parallel manipulators, (Liang and Ceccarelli 2012).

The reported previous solutions and experiences are the basis for the herein proposed kinematic design of a humanoid robot with a mechanism-based structure. In particular, in Fig. 7 the mayor mechanism for the proposed solution are illustrated as referring to: a LARM hand solution, Fig. 7a with only one actuator for the three fingers by suing a mechanical transmission with a common axle and a belt transmission; a waist/trunk module, Fig. 7b that is composed of a 3 d.o.f. parallel architecture with

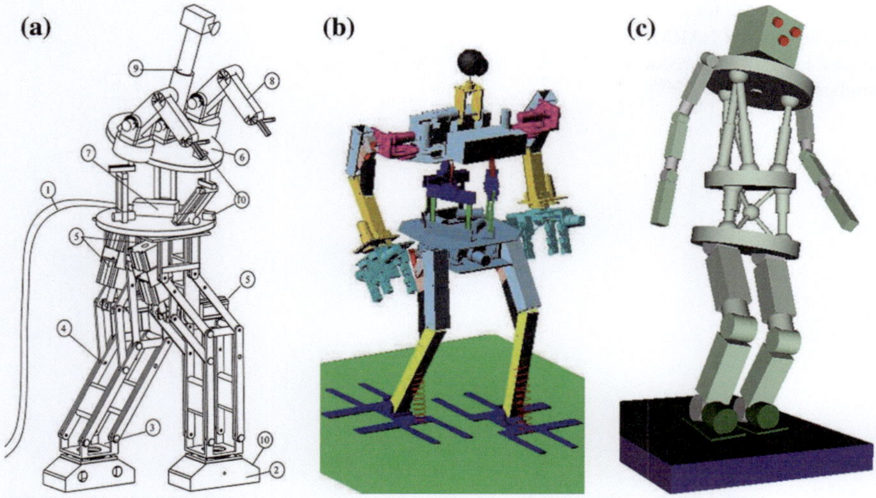

Fig. 6 LARM models for a mechanism-based humanoid robot with reduced number of d.o.f.s.: **a** linkage-based android, **b** CALUMA, **c** a humanoid robot with waist-trunk system with parallel manipulator architecture

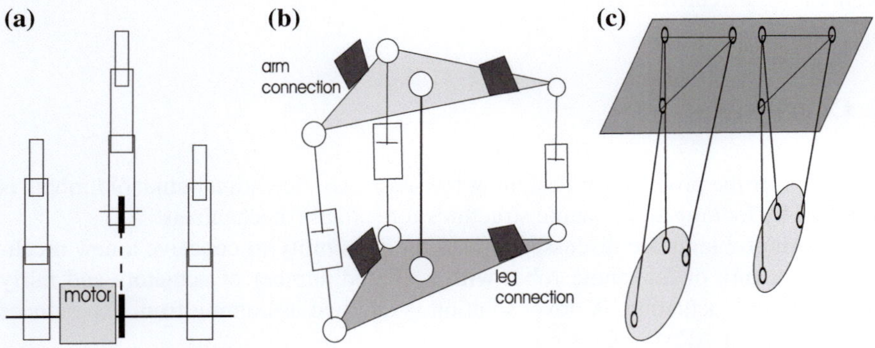

Fig. 7 Schemes of mechanism-based sub-systems for humanoid robot with a reduced number of d.o.f.s: **a** a three finger hand with one actuator, **b** a 3.do.f. trunk/waist module with parallel manipulator architecture, **c** a biped leg system with parallel manipulator architectures

a central bar and platforms as plates connecting with arms and bipedal system, as a module in a spine-like solution for multiple d.o.f.s mobility; a bipedal locomotion system, Fig. 7c, with two legs that are obtained by 3.d.o.f. parallel manipulators in which the mobile platforms are the feet moving with a pure translation with respect to the wais plate.

Combining all the above mechanisms a kinematic design of a mechanism-based humanoid robot can be proposed as in the sketch in Fig. 8. The peculiarity of this design can be recognized mainly in the possibility to fulfill the basic operation tasks

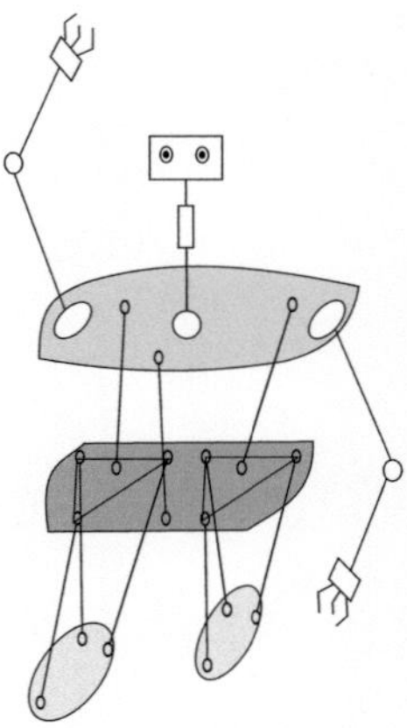

Fig. 8 A scheme for a mechanism-based LARM humanoid robot with reduced number d.o.f.s

in Tables 1 and 2 with a structure having the actuators in local mechanism frames for a fairly easy motion regulation and with a reduced number of actuated d.o.f.s for a light overall mechanical design with a considerable payload capability. In particular, the proposed design is obtained as based on linkages and parallel manipulators, as well known mechanism types.

4 Conclusions

In this paper the problem for designing low-cost easy-operation humanoid robots is attached by looking at kinematic structures for suitable mechanisms.

Design problems are discussed toward the possibility to conceive a new mechanism structure of humanoid robot with a limited number of actuators and fairly simple motion actuation. A novel solution is outlined as coming from experiences undergoing at LARM in Cassino.

References

Android World Homepage, Bipedal Projects (2013). http://www.androidworld.com/index.htm. Accepted 18 May 2013

Carbone G, Ceccarelli M (2008) Design of LARM hand: problems and solutions. J Control Eng Appl Inf 10(2):39–46

Carbone G, Ceccarelli M, Takanishi A, Lim H (2001) A study of feasibility for a low-cost humanoid robot. Humanoids, Tokyo, pp 351–358

Ceccarelli M, Carbone G (2009) A study of feasibility for a leg design with parallel mechanism architecture. In: IEEE/ASME conference on advanced intelligent mechatronics AIM'09. Singapore, pp 131

Gu H, Ceccarelli M, Carbone G (2010) Design and simulation of a 1 d.o.f. anthropomorphic clutched arm for robots. Int J Humanoid Rob 7(1):157–182. doi:10.1142/S0219843610002027

Honda Motor Co., Ltd (2013) ASIMO. http://world.honda.com/ASIMO/index.html. Accepted 18 May 2013

HRI Takanishi Lab (2013) Wabian. http://www.takanishi.mech.waseda.ac.jp/top/research/wabian/. Accepted 18 May 2013

JST (2013) PINO. http://orionrobots.co.uk/Pino. Accepted 18 May 2013

KAWADA Industries Inc (2013) HRP-4. http://global.kawada.jp/mechatronics/hrp4.html. Accepted 18 May 2013

Liang C, Ceccarelli M (2012) Design and simulation of a waist-trunk system for a humanoid robot. Mech Mach Theor 53:50–65

Li T, Ceccarelli M (2011) An experimental characterization of a rickshaw prop type. I J Mech Control 12(2): 29–48. ISSN 1590–8844

Nava Rodriguez NE, Carbone G, Ceccarelli M (2006) CaPaMan2bis as trunk module in CALUMA (CAssino low-cost hUMAnoid robot). In: 2nd IEEE international conference on robotics, automation and mechatronics RAM 2006, Bangkok, pp 347–352

Robotics A (2013) NAO. http://www.aldebaran-robotics.com/en/. Accepted 18 May 2013

Wu L, Carbone G, Ceccarelli M (2009) Designing an underactuated mechanism for a 1 active dof finger operation. Mech Mach Theor 44:336–348

Numerical Design Solutions for Telescopic Manipulators

Ericka Madrid Ruiz and Marco Ceccarelli

Abstract This paper proposes a design procedure for telescopic manipulators including a numerical solution when workspace is prescribed through few suitable points. An algebraic formulation for the workspace boundary is developed according to the analysis of the kinematic chain. Furthermore, a design algorithm is established by using those formulations and the numerical solution is worked out through a Newton-Raphson technique to solve a proper design problems. Numerical examples are reported to discuss computational efforts and solution characteristics.

Keywords Telescopic · Manipulator · Workspace · Numerical solutions · Design

1 Introduction

Most manipulators arms used in robotics are the telescopic arms since the prismatic pair as last joint gives to the kinematic chain some additional workspace capabilities. For that reason, the determination of the workspace volume and its characteristics has addressed and still address great attention in many works where the main purpose is to develop procedures for analyzing the workspace. According to that, there is very rich literature dealing with modeling, formulating, computing workspace characteristics with numerical procedures.

Dimensional design of manipulators is conveniently formulated by using workspace characteristics since manipulator reach ranges are recognized fundamental both for operation characterization and design purposes. Thus, in general algorithms for workspace determination have been used also in design procedures for reiterative analysis like those presented by Abdel-Malek and Yang (2006), Ceccarelli (1996a), Gupta and Roth (1982) and Tsai and Soni (1983) or for inverted formulation presented by Ceccarelli (1996b) and Freudenstein and Primrose (1984).

E. Madrid Ruiz (✉)
Pontifical Catholic University of Perú, Av. Universitaria 1801, Lima32, Perú
e-mail: madrid.ep@pucp.edu.pe

M. Ceccarelli
University of Cassino and South Latium, Via Di Biasio, 43 03043 Cassino, Italy
e-mail: ceccarelli@unicas.it

A. Kecskeméthy and F. Geu Flores (eds.), *Interdisciplinary Applications of Kinematics*, Mechanisms and Machine Science 26, DOI 10.1007/978-3-319-10723-3_11

In this paper the attention has been focused on the specific architecture of telescopic manipulators, with one prismatic joint at the end of the open kinematic chain with two revolute pairs. Practical application of telescopic manipulator architecture can be found not only in robotic arms, but also in many machines like for example excavators, firemen stairs, cleaning arms, concrete spraying equipment. Telescopic manipulator arms are widely used in robots since a prismatic pair as last joint in the chain gives additional workspace capability, but no great specific attention has been directed for the workspace analysis and design of this robot structure.

This paper is an attempt to extend the procedure presented by Ceccarelli (1996b) to the case of manipulator chains with prismatic pairs. In particular the formulation for the workspace of telescopic arms presented by Ceccarelli (1996a) has been expressed explicitly for design purposes. Preliminary results were presented by Ceccarelli (2012). In this paper a further development for design aims is proposed to discuss numerical solution of the design formulation and characteristics of design outputs as related to workspace prescriptions.

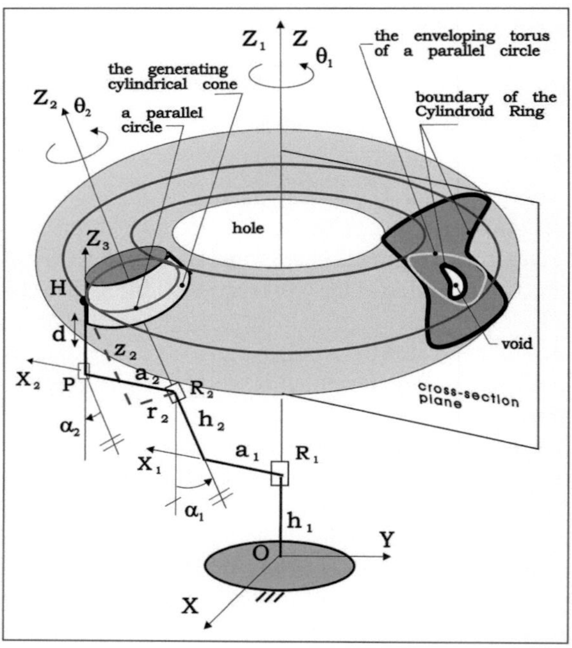

Fig. 1 Scheme for design parameters and workspace of telescopic manipulator arms

2 Kinematic Chain of a Telescopic Arm

A scheme for the kinematic chain of a telescopic arm is shown in Fig. 1 together with a geometric illustration of the workspace manifold and its generation. In particular, in Fig. 1 a world frame OXYZ has been assumed as fixed with the manipulator base with the Z axis coincident with the first joint axis and X axis as the reference axis for the first joint coordinates. Moving frames $O_i X_i Y_i Z_i$ $(i = 1, 2, 3)$ have been fixed on the links of the chain by assuming O_i in the center of the joint, Z_i as coinciding with the joint axis and X_i laying on the common normal between two consecutive joint axes. X_3 has been assumed to be parallel to X_2. The design HD parameters are the link lengths a_1 and a_2, the link offsets h_1 and h_2, the twist angles α_1 and α_2. However h_1 is considered meaningless since it does not affect the workspace but only shifts it up or down. The stroke excursion d is the design variable of the telescopic motion with the minimum and maximum values d_{min} and d_{max}, respectively. The time variable kinematic parameters are the joint angles θ_1 and θ_2 of the revolute pairs R_1 and R_2 and the joint stroke d of the sliding pair P. Each joint coordinate starts from a line which is parallel to X-axis of previous link. The joint angles θ_1 and $\theta 2$ are not considered as design variables since usually they vary for a full rotation and do not affect the workspace capability.

3 Determination of the Workspace

The workspace is generated by a reference point H on the extremity of the telescopic chain when H is moved to reach all the possible positions because of the movements of the joints. Workspace of telescopic manipulators can be characterized by looking at the figures which are generated by H because of successive movements of the joints starting from the last up to first which is fixed to the manipulator base, Fig. 1. Thus, movements of the sliding joint P generate a straight line segment which is limited by dmin and dmax. Then, the second revolute pair R_2 of the chain performs a full rotation of the straight line segment and generates a hyperboloid. Indeed, depending on the orientation of the straight-line segment with respect to the revolute joint axis Z_2 we may have a cylinder, a cone or generally a hyperboloid. Finally, a full revolution of the first revolute joint R_1 generates a solid of revolution with a general cross-section shape shown in Fig. 1. The workspace W(H) shows a hollow bulk shape and its cross-section is characterized by straight lines with possible cusps and two circular contours on the top and bottom.

The described procedure for workspace boundary generation can be synthesized in computing the workspace boundary as an envelope of the tori traced by mobility in R_1 and R_2 for the point of the segment due to P motion. In particular, each point of the straight line segment is individuated through a value of stroke d and the corresponding torus may be written by assuming $h_1 = 0$, with the hypothesis $\sin \alpha_1 \neq 0$ and with respect to OXYZ, by using the radial reach $r = (x^2 + y^2)^{1/2}$ and the axial reach z in the form

$$\left(r^2 + z^2 - A\right)^2 + (C\,z + D)^2 + B = 0 \tag{1}$$

in which the so-called structural coefficients are expressed as

$$A = a_1^2 + r_2^2 + (z_2 + h_2)^2\,;\; B = -4a_1^2 r_2^2;\; C = 2\frac{a_1}{s\alpha_1};\; D = -2a_1\,(z_2 + h_2)\,\frac{c\alpha_1}{s\alpha_1} \tag{2}$$

where r_2 and z_2 are the radial and axial reaches with respect to $O_2 X_2 Y_2 Z_2$. The distances r_2 and z_2 can be given from the geometry of the chain as

$$r_2 = \sqrt{a_2^2 + d^2 \sin^2 \alpha_2}; \quad z_2 = d \cos\alpha_2 \tag{3}$$

where the independent variable is the stroke parameter d.

Equations (1)–(3) can be used to determine the workspace volume for a given telescopic arm by scanning the stroke interval from a minimum value d_{min} to a maximum value d_{max}.

Indeed a torus is traced by using Eq. (1) to determine r by assuming z. This is obtained by scanning z within its range from minimum to maximum values, which can be calculated by using the model of Fig. 1 from

$$z = r_2 \sin\theta_2 \sin\alpha_1 + (h_2 + d\cos\alpha_2)\cos\alpha_1 + h_1 \tag{4}$$

Figure 2 shows a numerical example of the workspace determination for a given telescopic manipulator Eq. (7) in an concrete shot-machine by using the Eqs. (1)–(4).

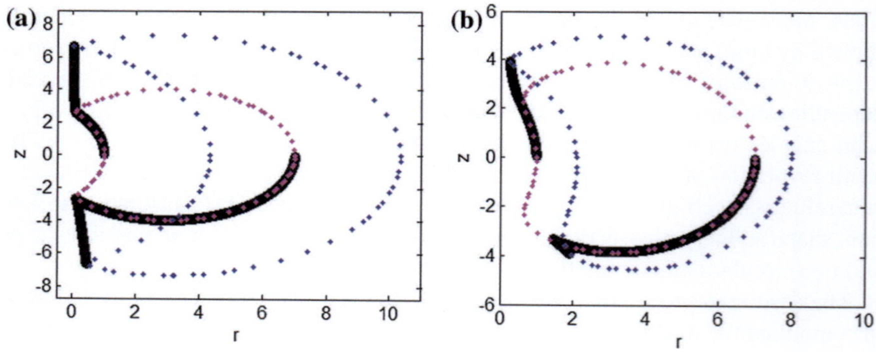

Fig. 2 Cross-section of the workspace boundary for a telescopic manipulator arm for spraying concrete (datasheet); **a** with a1 = 3, a2 = 4 u, h1 = 0, h2 = 0 u, α1 = 88 deg., α2 = 88 deg., dmin = 0 u, dmax = 6.15 u, **b** with α1 = 75 deg., α2 = 75 deg., dmax = 3.0 u. (u refers to an unit length). It is shown tori envelope segments (*thick black line*) and toroidal covers (*dotted points*)

Figure 2 shows that the workspace boundary is composed of two different geometrical topologies: envelope segments and toroidal surfaces. The envelope segments are located in the lateral sides of the cross-section representation, and two toroidal surfaces are the top and bottom covers, respectively. The envelope segments can be determined by a formulation by using the Theory of Envelopes to obtain a more efficient computational algorithm for workspace determination, Figs. 1 and 2. Thus, it is known that an analytical expression of the envelope can be obtained from Eq. (1) and its derivative with respect to the envelope parameter, which is the stroke d. Thus, differentiating Eq. (1) with respect to d we obtain

$$-\left(r^2 + z^2 - A\right) A' + (C\,z + D)\,D' + B' = 0 \tag{5}$$

where $'$ indicates d-derivative operator.

By using Eqs. (2) and (3) expressions for derivatives of the structural coefficients can be obtained in the form

$$A' = 2\,(d + h_2 \cos\alpha_2)\,;\; B' = -4a_1^2\, d \sin^2\alpha_2;\; D' = -2\, a_1 \cos\alpha_2 \frac{\cos\alpha_1}{\sin\alpha_1} \tag{6}$$

Finally, by using Eqs. (1)–(6) with suitable algebraic manipulations, the radial and axial reaches of points of envelope segments can be expressed with the hypothesis $A' \neq 0$ as

$$z = \frac{-B'D' \pm A'\sqrt{-\left[B'^2 + B\,\left(A'^2 + D'^2\right)\right]}}{\left(A'^2 + D'^2\right)\,C} - \frac{D}{C},\; r = \sqrt{\frac{B' + (C\,z + D)\,D'}{A'} + A - z^2} \tag{7}$$

The sign ambiguity exists to give two envelope segments.

The workspace determination of telescopic arms is completed with only two toroidal surfaces, namely the top and bottom covers. Figure 2 shows an example in which the covers of workspace have been determined by using upper branch of the top toroidal cover and lower branch of the bottom toroidal cover for an analysis also of the parameter influence.

4 Design Problems and Numerical Solutions

The design procedure is established by solving Eq. (7) as function of design parameters a_1, a_2, h_2, α_1, α_2. These equations are nonlinear algebraic equations giving the axial z and radial r reaches of points on the workspace envelope boundary as function of the design parameters of a RRP manipulator.

We have five unknowns as design parameters a_1, a_2, h_2, α_1, α_2 and an additional unknown as the telescopic stroke variable d. This can be used to identify, through the value d_i, each point z_i, r_i of the workspace envelope boundary so that you may write a solvable system of algebraic equations. The number n of design equations turns out to be equal to five in order to have as many equations as many unknowns. In addition the design of telescopic manipulator arms can be completed when telescopic stroke is calculated as limited within range limits d_{min} and d_{max}. These values can be determined by using the torus Eq. (1) for the top and bottom covers by expressing a torus as a function of d parameter.

Summarizing, a design algorithm for telescopic manipulators arms can be proposed when the workspace is prescribed through suitable seven boundary points by using the illustrated formulations in the following step-by-step procedure:

- solve Eq. (7) for five given workspace boundary points on the envelope branch to obtain the sizes a_1, a_2, h_2, α_1, α_2;
- solve torus equations Eqs. (1)–(3) for one given point on the top toroidal cover to calculate the stroke limit d_{max};

solve torus equations Eqs. (1)–(3) for one given point on the bottom toroidal cover to calculate the stroke limit d_{min}.

The choice and number of given workspace points are given by the conditions for the boundary envelope through five points and the toroidal covers (bottom and top) through two points. In general, these given points can be chosen at any positions within a space that is available with possible motion ranges of the manipulator joints. However, special configurations should be avoided as referring to degenerate topology for the workspace geometry that is described by the proposed Eq. (7) for the envelope boundary and proper Eq. (1) for the toroidal covers. In addition, the feasibility of solutions requires also considering constraints on the solvability of design Eqs. (7) and (1)–(3) in terms of real values for the design parameters. These problems are related to a numerical procedure that is explained by some numerical examples.

A numerical solution is approached by using Newton-Raphson technique to solve design equations as function of the unknowns vector x [a_1, a_2, h_2, α_1, α_2] as

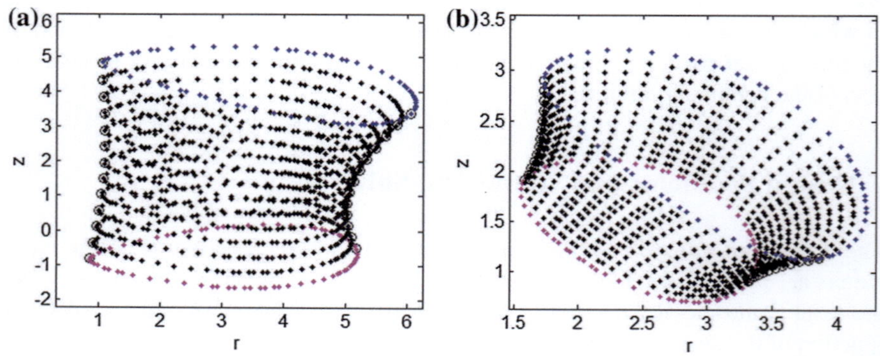

Fig. 3 **a** Cross section of the workspace boundary of the solution manipulator with $a1 = 3\,u$, $a2 = 2\,u$, $\alpha 1 = 25$ deg, $\alpha 2 = 25$ deg, $h2 = 1$, $d_{min} = -2\,u$, $d_{max} = 4\,u$ (u is length unit), **b** Cross section of the workspace boundary of the solution manipulator with $a1 = 2\,u$, $a2 = 1\,u$, $\alpha 1 = 45$ deg, $\alpha 2 = 45$ deg, $h2 = 2\,u$, $d_{min} = 0$, $d_{max} = 1.5\,u$ (u refers to an unit length)

Table 1 Numerical solutions for the cases of Fig. 3

Ex. (a)		a_1	a_2	α_1	α_2	h_2	*iter*
1	x_0	3	2	25	25	1	
		2.99977219	1.99949805	25.00518990	25.00293445	0.99972470	0
2	x_0	2	1	50	50	1	
		2.99977218	1.99949808	25.00519016	25.00293320	0.99972468	5
3	x_0	4	3	40	40	2	
		2.99977207	1.99949777	25.00519154	25.00294447	0.99972470	3
4	x_0	2	2	70	70	2	
		2.99977218	1.99949808	25.00519016	25.00293320	0.99972468	5
5	x_0	4	4	60	50	2	
		2.99977218	1.99949808	25.00519018	25.00293322	0.99972468	5
Ex. (b)		a_1	a_2	α_1	α_2	h_2	*iter*
1	x_0	2	1	45	45	2	
		1.99984371	1.00014895	44.99517098	44.96647295	1.99978865	0
2	x_0	1	1	30	30	2	
		1.99984407	1.00014937	44.99517122	44.96647605	1.99978891	3
3	x_0	1	1	50	50	1	
		1.99984409	1.00014938	44.99516999	44.96647440	1.99978888	3
4	x_0	4	4	60	60	3	
		1.99984409	1.00014938	44.99516999	44.96647439	1.99978888	4
5	x_0	3	3	70	70	3	
		1.99984409	1.00014938	44.99517000	44.96647438	1.99978887	5

$$f_i\,(\mathrm{x}) = \left\{\mathrm{r_i} - r_i^*;\ \mathrm{z_i} - z_i^*\right\} 0 \qquad i = 1, \ldots, 5 \tag{8}$$

where $\mathrm{r_i}$ and $\mathrm{z_i}$ are given by Eq. (7), and r_i^* and z_i^* are given as coordinates of the prescribed points of the boundary workspace of the configurations shown in Fig. 3.

A solution is efficiently reached after few iteration thanks to the algebraic formulation. The numerical solution computes the Jacobian by the finite differences method. This solution converges considering a tolerance of 1 e-6. In the numerical examples shown in Table 1, prescribed boundary workspace points have been selected in the envelope branch of a guess solution with, whose workspace is reported in Fig. 3. Two points have been indicated as extremity of the envelope branches so that they have been used also for determining the corresponding top and bottom torus covers.

Table 1 list solutions of two examples that converge to a manipulator design with workspace of Fig. 3, even though different and far guess manipulators have been considered. This is because the prescribed points can be satisfied by that manipulator that is nevertheless obtained even form far guess solution. Thus the proposed formulation and corresponding numerical design procedure can provide a design solution to a feasible manipulator with very accurate fast convergence starting from any guess solution. However, during numerical test experiences it has been noted that convergence is not obtained for the cases when the design parameters α_1 and α_2 have values near to zero or 90 deg, which are the conditions of singularity for Eq. (7).

5 Conclusions

In this paper the analytical formulation proposed for the workspace boundary allows to solve numerically the design problem for telescopic manipulator arms. In particular the workspace boundary has been formulated as composed by an envelope branch and two toroidal covers. The envelope branch has been formulated by using an envelope process with a suitable family of toroidal surfaces. The same torus equation has been used to determine the top and bottom covers. These formulations have been used to express the design problem of telescopic manipulator arms when the workspace is prescribed by seven boundary points. A numerical procedure has been elaborated to solve the design problem with a suitable number of prescribed boundary workspace points matching the number of available equations. A numerical example is used to outline convenient computational efforts and feasibility of solutions.

Acknowledgments The authors wish to acknowledge the support of the Pontifical Catholic University of Peru in Lima during the sabbatical stay of prof. M. Ceccarelli in fall semester of 2012.

References

Abdel-Malek K, Yang J (2006) Workspace boundaries of serial manipulators using manifold stratification. Int J Adv Manuf Technol 28:1211–1229

Ceccarelli M (1996a) A workspace analysis for RRP manipulators. In: CD Proceedings of 1996 ASME 24th biennial mechanisms conference, paper 96DETC-Mech1012

Ceccarelli M (1996b) Formulation for the workspace boundary of general N-revolute manipulators. Mech Mach Theor 31:637–646

Ceccarelli M (2012) A formulation for analytical design of telescopic manipulators with prescribed workspace. Appl Mech Mater 162:113–120

Freudenstein F, Primrose EJF (1984) On the analysis and synthesis of the workspace of a three-link, turning-pair connected robot arm. ASME. J. Mech Trans Autom Design 106:365–370

Gupta KC, Roth B (1982) Design considerations for manipulator workspaces. ASME J Mech Design 104:704–711

SPM5314 (2013) Datasheet for Telescopic spraying arm SPM5314. http://shotcrete.putzmeister.es/audiovisual/catalogos/5314_E.pdf. Accepted 31 May 2013

Tsai YC, Soni AH (1983) An algorithm for the workspace of a general n-R robot. ASME J Mech Trans Autom Design 105:52–57

A Sequentially-Defined Kinetostatic Model of the Knee with Anatomical Surfaces

Irene Sintini, Nicola Sancisi and Vincenzo Parenti-Castelli

Abstract Recently, a static model of the knee featuring a spherical approximation of the contact surfaces has been defined by a sequential approach, in order to generalize a kinematic model of the joint passive motion. The aim of this study is to define a more sophisticated static knee model with articular surfaces which accurately reproduce the anatomical ones. The motion of the new model is compared, under several loading conditions, with the motion of the knee when the same loads are applied and with the motion of the previous static model. Although some improvements could be achieved by a proper optimization of the model parameters, the model is able to replicate the behavior of the human knee under both passive and loaded conditions.

1 Introduction

The knee joint has always played a central role in biomechanic studies for its dual function of stabilizing and enabling a wide range of motion. To investigate both the kinematic and the static knee behavior, many mathematical models have been presented in the literature (Hefzy and Cooke 1996) and two approaches can be identified: a simultaneous one, whose purpose is to solve the kinematic and the dynamic analyses of the system at the same time, and a sequential one, which first solves the kinematic

I. Sintini (✉) · N. Sancisi
DIN - Department of Industrial Engineering, University of Bologna, Bologna, Italy
e-mail: irene.sintini2@unibo.it

N. Sancisi
e-mail: nicola.sancisi@unibo.it

V. Parenti-Castelli
DIN - Department of Industrial Engineering, Health Sciences and Technologies,
Interdepartmental Centre for Industrial Research (HST-ICIR),
University of Bologna, Bologna, Italy
e-mail: vincenzo.parenti@unibo.it

A. Kecskeméthy and F. Geu Flores (eds.), *Interdisciplinary Applications of Kinematics*,
Mechanisms and Machine Science 26, DOI 10.1007/978-3-319-10723-3_12

analysis and then, on the basis of these results, solves the static and the dynamic analyses. A first sequentially-defined model has been recently presented (Sancisi and Parenti-Castelli 2011): the characteristics of the sequential approach make the model able to replicate both the passive motion, i.e. the motion when no loads are applied, and the loaded motion (Franci et al. 2009; Sancisi and Parenti-Castelli 2011).

This paper further investigates the sequential procedure potentialities, by the development of a new static model which differs from the previous one (Sancisi and Parenti-Castelli 2011) for a more accurate reproduction of the ligaments and of the natural articular surfaces, which were approximated by spheres in the previous model. The final aim is to understand to which extent the model results are improved by a better approximation of the anatomical structures. The new model was tested under several loading conditions and the results were then compared to the ones obtained from the previous static model, and to experimental data from the literature.

2 Methods

The sequential approach used in this paper for the static model definition consists of several steps (Franci et al. 2009; Sancisi and Parenti-Castelli 2011). First, a kinematic model of the knee (M1) featuring the joint main passive structures is defined in order to accurately reproduce the passive motion of this joint. Afterward, a static model (M2) is defined as a generalization of M1, by adding the other passive structures and by considering the viscoelastic properties of all included structures. M2 can replicate the joint behavior when only the external loads, i.e. no muscle loads, are applied. This model is devised so that the results of M1 are not worsened by generalization: M1 parameters are not changed during M2 definition and the new added structures do not alter the results of M1. This means that M2 can replicate the knee passive motion as well as M1, because all the ligamentous structures existing in M2 but not in M1 remain slack during this motion by a proper identification of the M2 model parameters.

2.1 Kinematic Model M1 and Previous Static Model M2

The kinematic model M1 is based on an equivalent mechanism presented in previous studies (Parenti-Castelli and Gregorio 2000; Sancisi and Parenti-Castelli 2010; Ottoboni et al. 2010; Parenti-Castelli and Sancisi 2013). The aim of M1 is to replicate the passive motion of the knee, which was shown to be guided by articular contacts and by three ligamentous isometric fibers (Wilson et al. 1998): one fiber of the anterior cruciate ligament (ACL), one of the posterior cruciate ligament (PCL) and one of the medial collateral ligament (MCL). At this step, the knee is modeled as a one degree-of-freedom mechanism (Fig. 1a) composed of two rigid bodies, the femur and the tibia, connected by five rigid links: three of them are the isometric fibers and the

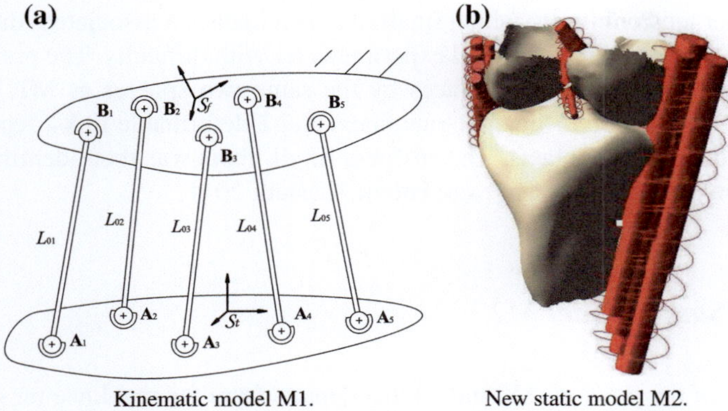

Fig. 1 Kinematic model M1 and its generalization, i.e. the static model M2 of the knee. In (*b*) the ligaments (depicted as springs) and the anatomical surfaces are represented. **a** Kinematic model M1, **b** New static model M2

remaining two represent the contacts between the lateral and medial condyles, since the articular surfaces were approximated by spheres whose centers always maintain the same mutual distance (Parenti-Castelli and Gregorio 2000). Once the geometrical parameters of the model, i.e. the isometric lengths L_{0i} and the insertion coordinates $\mathbf{A}_i$ on the tibia and $\mathbf{B}_i$ on the femur of each link, are obtained from specimen, the passive motion can be obtained by solving the mechanism closure equations. The five closure equations can be written by imposing the relevant fixed length L_{0i} for each rigid link i (Parenti-Castelli and Gregorio 2000; Parenti-Castelli and Sancisi 2013). The procedure of identification based on optimization is described in previous studies (Sancisi and Parenti-Castelli 2010; Parenti-Castelli and Sancisi 2013).

After obtaining M1, the second step of the sequential approach allows the definition of the static model M2 as a generalization of M1. Indeed, as previously reported above, a first static model M2 (Sancisi and Parenti-Castelli 2011) was already defined by the sequential procedure as follows. The main knee ligamentous structures were incorporated in the model, namely 3 isometric fibers plus 21 new fibers, and all the fibers were considered as elastic structures: it is worth noting that also the isometric fibers are considered deformable in M2, but they are still called isometric hereafter to differentiate them from the others. Preliminary estimates of stiffness and geometrical parameters of the new 21 fibers, as well as the stiffness of the isometric fibers, were established based on published and experimental data. The insertion points and the zero-load lengths of the isometric fibers were the same as M1, as required by the sequential approach. As regards the new fibers, they must be slack or just slightly tensioned during the passive motion in order to respect the sequential approach's philosophy, hence their zero-load lengths were forced to be greater than the maximum distances between the insertion points during the passive flexion arc. This inferior bound for the new fibers is then reduced by 1 %: even if, strictly speaking, this violates the rules of the sequential approach, this small reduction simulates possible fiber tensionings (to which small ligament loads are associated) during passive motion that could be measured experimentally with difficulty. The previous M2 approximated the anatomical surfaces by the same two spheres as M1, therefore the model consisted of 24 ligamentous fibers and 2 deformable links, representing the sphere-on-sphere contacts. The stiffness of all fibers was then identified by an optimization procedure (Sancisi and Parenti-Castelli 2011).

2.2 New Static Model M2

The feature of the previous M2 model that departs from the real knee most was the spherical approximation of the articular surfaces. The aim of this new M2 model is to improve this aspect. Hence the new M2 presents an accurate anatomical reproduction of the knee articular surfaces; moreover, it features additional ligaments (previously not included) and a different force-strain relation for the elastic fibers, in order to better replicate the anatomy of the knee. Strictly speaking, the M1 model corresponding to the new M2 model is no longer the mechanism in Fig. 1a: it is kinematically equivalent to it, thanks to the procedure used to generate the surfaces described in Sect. 2.2.1, but it would provide different results in terms of contact paths and articular forces, since the contact surfaces are no longer spheres.

2.2.1 Articular Surfaces

The femoral and tibial articular surfaces are measured on the same specimen used for M1 by a stereophotogrammetric system, which records the surfaces as point clouds, and by CT scans that are used to obtain a 3D model of the bones by segmentation. Anatomical reference points are measured to define two anatomical coordinate systems, $\mathscr{S}_f$ for the femur and $\mathscr{S}_t$ for the tibia (Sancisi and Parenti-Castelli 2010; Ottoboni et al. 2010). The axes of $\mathscr{S}_f$ and $\mathscr{S}_t$ are oriented along the antero-posterior, proximo-distal and medio-lateral directions; the origin of $\mathscr{S}_f$ is the midpoint between the femur epicondyles, while the origin of $\mathscr{S}_t$ is the center of the tibia plateau. The femoral surface point clouds are aligned to the femur 3D model from CT data, in order to combine the accuracy of the first method to the completeness of the second one; Rhinoceros 3D (Robert McNeel and Associates, Seattle, WA) is used for the alignment. The same procedure is repeated on the tibia. However, the sequential procedure requires that the new contact surfaces do not alter the passive motion of the knee: the constraints provided by the anatomical surfaces must be kinematically equivalent to the two sphere-on-sphere contacts in M1. The envelope procedure is used for this scope: the femoral aligned surfaces are moved by the M1 model in order to obtain their conjugated surfaces on the tibia during a passive flexion arc; two relevant sections of these conjugated surfaces are cut out around the femur and tibia contact areas during passive motion. Then, these sections are merged with the aligned experimental surfaces of the tibia. The final tibia surfaces are very similar to the experimental ones, thanks to the accuracy of M1. The final meshes are exported to Adams (MSC Software Corporation, Santa Ana, CA), where the new M2 model is implemented (Fig. 1b). Rigid contact is imposed between the tibial and femoral articular surfaces.

2.2.2 Ligament Properties

Thirty-five fibers are used to model the main knee ligamentous structures (the number of fibers is in brackets). Some ligaments were already considered in the previous M2: ACL (4+1 isometric), PCL (4+1 isometric), MCL (5+1 isometric), LCL (3), oblique popliteus ligament (3), popliteus tendon (2). Some posterior ligaments are added to the previous M2: the articular capsule (3) (Rachmat et al. 2012), the posterior oblique ligament (2) (Prade et al. 2003, 2005; Robinson et al. 2005; Hughston and Eilers 1973), the fabellofibular ligament (2) (Diamantopoulos et al. 2005; Minowa et al. 2004), the midthird lateral capsular ligament (4) (Terry and Prade 1996). Geometrical and mechanical parameters, i.e. fiber insertion points coordinates, zero-load lengths and fiber stiffness, are the same as the preliminary estimates used in the previous M2 if the fiber is common to both models: only few adjustments are made on the stiffness and on some insertion points based on experimental data in the literature. As for the fibers added to the new model, the parameters are estimated according to the data available in the literature (Prade et al. 2003, 2005; Diamantopoulos et al. 2005; Robinson et al. 2005; Hughston and Eilers 1973; Terry and Prade 1996) and the zero-load lengths are calculated in order to respect the sequential approach described in Sect. 2.1. Then, the zero-load lengths of new structures are reduced by 1 % like it was done in the previous M2.

The force-displacement curve for the ligaments is assumed to be parabolic-linear (Blankevoort et al. 1991; Bloemker et al. 2012):

$$\begin{aligned} F &= \frac{1}{4}k\frac{\varepsilon^2}{\varepsilon_l} \quad 0 < \varepsilon \leq 2\varepsilon_l \\ F &= k(\varepsilon - \varepsilon_l) \quad \varepsilon > 2\varepsilon_l \\ F &= 0 \quad \varepsilon \leq 0 \end{aligned} \tag{1}$$

In Eq. (1) k is the stiffness in Newton and ε is the strain of the fiber defined as $\varepsilon = \frac{L-L_0}{L_0}$ where L and L_0 are respectively the length and the zero-load length of the fiber; ε_l is assumed to be 0.03 (Bloemker et al. 2012). In the previous M2, the force-displacement curve was parabolic (Sancisi and Parenti-Castelli 2011): in the new M2 the stiffness parameters of old fibers are modified in order to fit the parabolic-linear curve. Since the linear portion of Eq. (1) plays a role only for high loads, the effect of the shift from a parabolic force-displacement curve to a parabolic-linear one is very small.

To implement the model in Adams, each fiber is modeled as a spring (Fig. 1b) whose force-displacement behavior described in Eq. (1) is imposed by a user-written subroutine (Bloemker et al. 2012); a damping coefficient of 1 Ns/mm is put in parallel with each fiber, to reach the static equilibrium faster. Contrary to the previous M2, all the parameters of fibers added to M1 are not optimized in this case and thus their final value is chosen as noted above, apart from very few manual adjustments congruent with the reference literature.

2.2.3 Loading Conditions and Kinematic Measurements

As the previous static model M2 (Sancisi and Parenti-Castelli 2011), the new model is tested under several clinically significant loading conditions, namely anterior and posterior drawer tests, internal and external torsion tests, ab/adduction tests. The relative motion between the femur and tibia under each different loading condition is obtained in terms of joint angles (Grood and Suntay 1983) and it is then compared to the results presented in Grood et al. (1988), which are chosen as a reference, and to the results computed by the previous M2 model (Sancisi and Parenti-Castelli 2011). To be consistent with the testing conditions of Grood et al. (1988), the femur is fixed and the tibia is allowed to move freely under the effect of its own weight (50 N) and of external forces ($\pm$5 Nm for the torsion tests, $\pm$20 Nm for the ab/adduction tests, $\pm$100 N for the drawer tests). The flexion is blocked during each test, namely the model has 5 degrees of freedom: in Grood et al. (1988) the tibia was not allowed to rotate around the medio-lateral axis when the loads were applied, since the aim of the Authors was to investigate the relative motion between the tibia and femur at several flexion angles.

3 Results

The new static model M2 allows the reproduction of both the passive and the loaded motion of the knee with an accuracy comparable to the previous M2 model. Each plot in Fig. 2 presents the results of the new model, of the previous model and of the experimental reference data. Figure 2a shows the results of the drawer test: higher and lower curves respectively represent the anterior and posterior displacements of the tibia with respect to the femur, taking as a reference for each flexion angle the corresponding relative pose when only tibia weight is applied (Grood et al. 1988). Likewise, Fig. 2b, c show respectively the results of the torsion and of the ab/adduction tests: lower curves represent the internal rotation and the abduction and higher curves represent the external rotation and the adduction of the knee. The middle curves in Fig. 2b, c show the motion of the joint when only tibia weight is applied: the pose of the knee at full extension during this motion is the reference pose for all the rotation measurements (Grood et al. 1988).

The mean absolute errors between experimental and simulated motion (expressed as percent values with respect to the maximum range of the corresponding motion) are 15.0 % for the drawer, 7.7 % for the internal/external rotation, 17.8 % for the

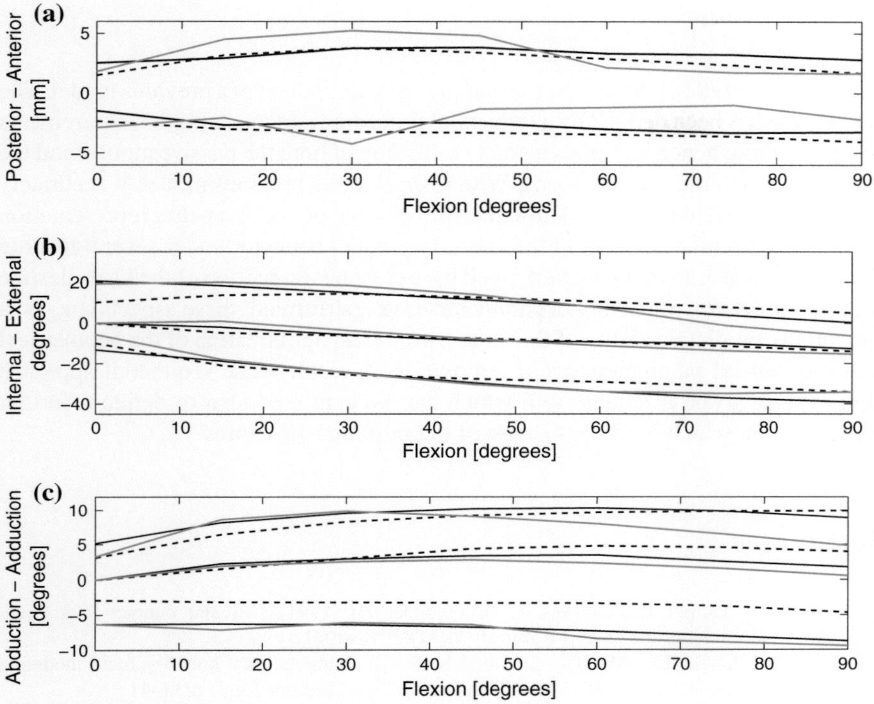

Fig. 2 Results of the new M2 model (*solid*, *gray*) compared to the previous M2 model results (*solid*, *black*) and to the experimental data (Grood et al. 1988) (*dash*). **a** Drawer Test, **b** Torsion Test, **c** Ab/adduction Test

ab/adduction tests. As for the torsion tests, the previous and the new M2 model behave similarly. Both models present the highest errors in the abduction test, since they appear to be less rigid than the reference knee. The new M2 is also stiffer than the reference knee in the adduction test at high flexion angles, but, in this case, the previous model presented better results than the new one. However, as previously noted, unlike the previous M2, no systematic optimization is performed in the new M2, but the stiffness and the zero load lengths of some fibers have been slightly adjusted. It can be inferred that the accuracy of the new M2 could be improved through a proper optimization of the mechanical and geometrical parameters of the ligaments. In addition, the results could be improved also by defining a compliant contact between femoral and tibial articular surfaces instead of a rigid one, but previous investigations revealed a low influence of the contact stiffness on the knee behavior (Blankevoort et al. 1991). Finally, it is worth noting that the reference results for the static model are average values from the literature (Grood et al. 1988): the motion of the considered specimen could be actually different from the reference one. Thus, a stronger validation will be required with static data measured on the same specimen.

4 Conclusions

A new static model of the knee is presented as a development of a previous model. The new model has been defined by a sequential procedure and meets all the requirements of this method, hence it ensures a good replication of both the passive motion and the motion under static external loads. With respect to the previous model, it accurately reproduces the real anatomical articular surfaces and provides a better representation of all ligamentous structures in the knee. The model behavior under several loading conditions proves to replicate fairly well the experimental motion of the knee, despite a full optimization of the model parameters is not performed: these aspects suggest a potentially predictive ability of the model, even if an optimization of the geometrical and mechanical parameters could improve these results. The sequential approach proves itself to be a reliable tool which can be exploited also to define a further development, which involves the role of the muscular structures.

References

Blankevoort L, Kuiper JH, Huiskes R, Grootenboer HJ (1991) Articular contact in a three-dimensional model of the knee. J Biomech 24(11):1019–1031

Bloemker KH, Guess TM, Maletsky L, Dodd K (2012) Computational knee ligament modeling using experimentally determined zero-load lengths. Open Biomed Eng J 6:33–41

Diamantopoulos A, Tokis A, Tzurbakis M, Patsopoulos I, Georgoulis A (2005) The posterolateral corner of the knee: evaluation under microsurgical dissection. J Arthrostopi Relat Surg 21(7): 826–833

Franci R, Parenti-Castelli V, Sancisi N (2009) A three-step procedure for the modelling of human diarthrodial joints. Int J Mech Control 10(1):3–12

Grood ES, Stowers SF, Noyes FR (1988) Limits of movement in the human knee. J Bone Joint Surg 70-A(1):88–97

Grood ES, Suntay WJ (1983) A joint coordinate system for the clinical description of three-dimensional motions: application to the knee. J Biomech Eng 105:136–144

Hefzy MS, Cooke TDV (1996) Review of knee models: 1996 update. Appl Mech Rev 49 (10–2):187–193

Hughston JC, Eilers AF (1973) The role of the posterior oblique ligament in repairs of acute medial (collateral) ligament tears of the knee. J Bone Joint Surg [Am] 55(5):923–940

La Prade RF, Bollom TS, Wentorf FA, Wills NJ, Meister K (2005) Mechanical properties of the posterolateral structures of the knee. Am J Sports Med 33(9):1386–1391

La Prade RF, Ly TV, Wentorf FA, Engebretsen L (2003) The posterolateral attachments of the knee. Am J Sports Med 31(6):854–860

Minowa T, Murakami G, Kura H, Suzuki D, Han SH, Yamashita T (2004) Does the fabella contribute to the reinforcement of the posterolateral corner of the knee by inducing the development of associated ligaments? J Orthop Sci 9:59–65

Ottoboni A, Parenti-Castelli V, Sancisi N, Belvedere C, Leardini A (2010) Articular surface approximation in equivalent spatial parallel mechanism models of the human knee joint: an experiment-based assessment. Proc Inst Mech Eng Part H J Eng Med 224(9):1121–1132

Parenti-Castelli V, Di Gregorio R (2000) Parallel mechanisms applied to the human knee passive motion simulation. In: Lenarcic J, Stanisic M (eds) Advances in robot kinematics. Kluwer Academic, Boca Raton

Parenti-Castelli V, Sancisi N (2013) Synthesis of spatial mechanisms to model human joints. In: McCarthy M (ed) 21th century kinematics. Springer, Berlin, pp 49–84

Rachmat HH, Janssen D, van Tienen T, Diercks RL, Verkerke B, Verdonschot N (2012) Material properties of the human posterior knee capsule. J Biomech 45(S1):S380

Robinson JR, Bull AMJ, Amis AA (2005) Structural properties of the medial collateral ligament complex of the human knee. J Biomech 38:1067–1074

Sancisi N, Parenti-Castelli V (2010) A 1-dof parallel spherical wrist for the modelling of the knee passive motion. Mech Mach Theor 45:658–665

Sancisi N, Parenti-Castelli V (2011) A sequentially-defined stiffness model of the knee. Mech Mach Theor 46:1920–1928

Terry GC, La Prade RF (1996) The posterolateral aspect of the knee. Anatomy and surgical approach. Am J Sports Med 24(6):732–739

Wilson DR, Feikes JD, O'Connor JJ (1998) Ligaments and articular contact guide passive knee flexion. J Biomech 31(12):1127–1136

Designing and Implementing an Autonomous Navigation System Based on Extended Kalman Filter in a CoroBot Mobile

Gerardo Arturo Vilcahuamán Espinoza and Edilberto Vásquez Díaz

Abstract Autonomous mobile robots are a common trend in robotics these days. They characterize for combining their mobile abilities and perception capacities in an intelligent way. This paper explains the work done to achieve two basic issues in autonomous navigation: position tracking (localization) and path planning. It starts with an odometric model of robot as an approximation to a differential drive mobile, including probabilistic modeling for uncertainty evolution. Next, based on the Extended Kalman Filter, it continues to show a form to integrate odometric predictions with sensor data, so that pose belief could pass through measurement update. Once the mobile position is tracked, robot is able to follow a trajectory towards a specific position objective, avoiding roadblocks in the way. This trajectory would be computed using a "Base Points" method for path planning. Finally, results of the implementation of the navigation system are presented.

1 Odometric Model for CoroBot

A model based on Odometry is developed from kinematic constraints to probabilistic approach, taking into consideration CoroBot mobile limitations.

CoroBot, made by CoroWare®, is a mobile platform designed for robotics investigation. It consists on a four-wheeled vehicle with a rectangular chassis. Its wheels are electrically connected two by two (right and left), so that each side has independent control.

The robot mobility will be simplified as the movement of a wheeled rigid body on a plane horizontal surface. Therefore, **pose** (or kinematic state) would be represented by $x_t = [x \quad y \quad \theta]^T$, where x, y stand for planar position and θ stands for heading direction. The relationship between local and global velocities (V_L and

G.A. Vilcahuamán Espinoza (✉) · E. Vásquez Díaz
Department of Electronics and Automation, University of Piura, Piura, Peru
e-mail: gerardo.ve90@gmail.com

E. Vásquez Díaz
e-mail: edilberto.vasquez@udep.pe

A. Kecskeméthy and F. Geu Flores (eds.), *Interdisciplinary Applications of Kinematics*, Mechanisms and Machine Science 26, DOI 10.1007/978-3-319-10723-3_13

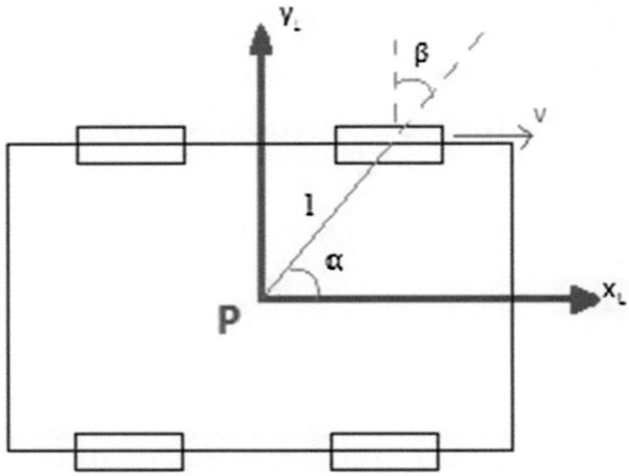

Fig. 1 Skid steering mobile with local reference system

V_G, respectively) can be settled using a rotation matrix as axis for local reference is aligned to the actual heading: $V_L = R(\theta)V_G$.

In order to obtain a kinematic model of the robot, it is necessary to state the two kinematic constraints of each of its four standard fixed wheels. The first one is pure rotation, which means that the amount of movement in its plane will always be comparable to the correspondent spin, as no slip appears. The second one is zero lateral motion, which states that no wheel would slip in a direction orthogonal to its plane. Both constraints are represented by the following equations, respectively.

$$[\sin(\alpha+\beta) \quad -\cos(\alpha+\beta) \quad -l\cos\beta]R(\theta)V_G - r\dot{\varphi} = 0 \tag{1}$$

$$[\cos(\alpha+\beta) \quad \sin(\alpha+\beta) \quad l\sin\beta]R(\theta)V_G = 0 \tag{2}$$

Since CoroBot is a skid steering mobile (it has four standard fixed wheels, with independent control for each side), the kinematic model is obtained from its wheel kinematic constraints as shown bellow. Figure 1 presents the variables used for modeling.

$$\begin{bmatrix} 1 & 0 & -l\cos\beta \\ 1 & 0 & l\cos\beta \\ 0 & 1 & l\sin\beta \\ 0 & 1 & -l\sin\beta \end{bmatrix} R(\theta)V_G - \begin{bmatrix} r\dot{\varphi}_1 \\ r\dot{\varphi}_2 \\ 0 \\ 0 \end{bmatrix} = 0 \tag{3}$$

where $\dot{\varphi}_1, \dot{\varphi}_2$ are both wheel angular speeds, and r is the wheel radius.

Equation (3) proves that it is not possible to obtain global velocity from angular speeds, since first matrix cannot be inverted. The only case when this is possible is when both angular speeds have the same value (the mobile goes forward or backward in a straight line). Any other way of motion will necessarily cause slip, because it would break kinematic constraints as Eq. (3) won't have a solution for V_G .

Although skid steering systems present difficulties for kinematic modeling like the one shown before, in this case it is possible to avoid more complex control approaches (Kozlowski and Pazderski 2004) if we approximate them to a differential scheme instead. For this purpose, we shall make two considerations:

- Motion in straight line (forward and backward) can be modeled the same way, either for differential or skid steering mobiles
- When opposite angular speeds were applied on wheels, there was little slip as the robot almost spun around its own axis. This slip could be treated with correction factors in a differential model.

If Eqs. (1) and (2) are applied to the mobile but considering a differential architecture this time, the odometric model could be achieved. The actual pose x_t would be a function of previous pose x_{t-1} and movement generated from wheel spins in the last period of time.

$$x_t \approx \begin{bmatrix} x \\ y \\ \theta \end{bmatrix}_{t-1} + \begin{bmatrix} \frac{\Delta s_2+\Delta s_1}{2} . \cos\left(\theta + \frac{\Delta s_2-\Delta s_1}{2b}\right) \\ \frac{\Delta s_2+\Delta s_1}{2} . \sin\left(\theta + \frac{\Delta s_2-\Delta s_1}{2b}\right) \\ \frac{\Delta s_2-\Delta s_1}{b} \end{bmatrix} \tag{4}$$

where Δs_1, Δs_2 are the angular displacement for each wheel and b is the distance between wheels.

Despite having achieved the odometric equation, the fact is that motion uncertainty will increase as the robot advances causing that pose estimation will be less accurate each time. This is known as error propagation. Odometric measurements have accumulative errors that generate miscalculations of future poses. This uncertainty can be modeled by a covariance matrix, considering that pose behaves as a Gaussian function. This covariance will depend on former covariance and motion covariance (which represents the uncertainty caused from motion) (Siegwart and Nourbakhsh 2004).

$$\overline{\Sigma}_t = G_t \Sigma_{t-1} G_t^T + G_\Delta \Sigma_\Delta G_\Delta^T \tag{5}$$

where Jacobians are defined as: $G_t = \left[\frac{\partial f}{\partial x} \quad \frac{\partial f}{\partial y} \quad \frac{\partial f}{\partial \theta}\right]$ and $G_\Delta = \left[\frac{\partial f}{\partial \Delta s_1} \quad \frac{\partial f}{\partial \Delta s_2}\right]$.

2 EKF Localization

As an extended approach of Kalman Filter, EKF permits state updates in non-linear systems using Taylor approximations for Gaussian systems.

State shall be considered as the collection of all robot and environment features that will impact the future. In this case, pose x_t is the state that is being analyzed.

In probabilistic robotics there are two main factors that influence pose: environmental sensor measurement (z_t) and control action (u_t). When state is treated as a probabilistic distribution, it can also be referred as **belief** (Thrun 2002).

Kalman Filter is the most popular technique among Gaussian filters. However, belief computations are limited to continuous spaces only. KF requires that a posterior distribution fulfills three conditions:

1. State transition probability $\boldsymbol{p(x_t|u_t, x_{t-1})}$ shall be a linear function with Gaussian noise.
2. Measure probability $\boldsymbol{p(z_t|x_t)}$ shall be a linear function with Gaussian noise.
3. Initial belief $\boldsymbol{bel(x_0)}$ shall follow a normal distribution.

In this methodology, the measurement update step is accomplished using a new factor: **Kalman gain**(K_t). This value will represent the degree at which sensor data is incorporated to the new state estimation. Notice that while KF works with Gaussian distributions, Kalman gain will be used to update probabilistic parameters for multivariable normal distributions, which are the mean μ_t and the covariance matrix Σ_t.

Since almost all state transitions and measures in environment are non-linear, pose belief would not prevail as a Gaussian distribution (this is because a random Gaussian variable can be obtained from another random Gaussian variable only if this last goes through a linear transformation). Extended Kalman Filter considers state transition probability, $x_t = g\ (u_t, x_{t-1}) + \varepsilon_t$, and measure probability, $z_t = h\ (x_t) + \delta_t$, as non-linear functions, just like they usually are in reality. However, instead of computing the exact posterior, EKF approximates mean and covariance of the new distribution by linearizing them through first order Taylor expansion (Suliman et al. 2009).

$^{\alpha}G_t$, H_t are Jacobians from g, h functions respectively; R_t is the motion covariance matrix from g function; and Q_t is the sensor covariance matrix from h function (Thrun et al. 2006).

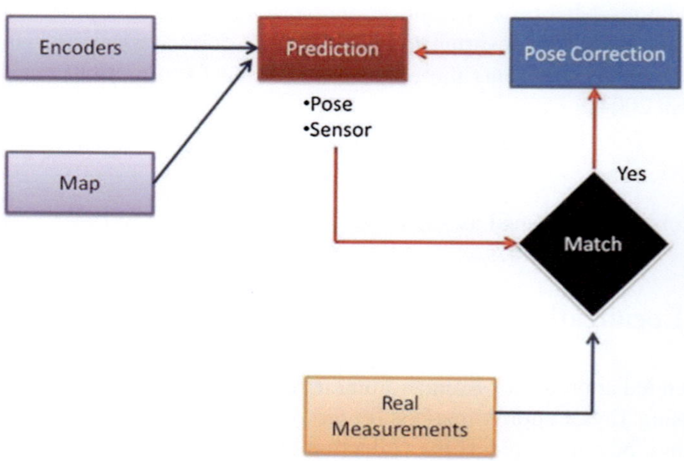

Fig. 2 EKF localization diagram

Table 1 Extended Kalman filter algorithm

$EKF\_algorithm^{\alpha}\ (\mu_{t-1}, \Sigma_{t-1}, u_t, z_t)$:
$\overline{\mu}_t = g\ (u_t, \mu_{t-1})$
$\overline{\Sigma}_t = G_t \Sigma_{t-1} G_t^T + R_t$
$K_t = \overline{\Sigma}_t H_t^T \left(H_t \overline{\Sigma}_t H_t^T + Q_t\right)^{-1}$
$\mu_t = \overline{\mu}_t + K_t \left(z_t - h\left(\overline{\mu}_t\right)\right)$
$\Sigma_t = (I - K_t H_t) \overline{\Sigma}_t$
Return μ_t, Σ_t

EKF Localization aims to update pose belief from time $t-1$ to time t using control data and sensor measures, as well as map features. Since robot will part from a known initial pose, and will be tracing probabilistic evolution of its own pose in time, it would be performing **position tracking** (Fig. 2).

EKF Localization utilizes map data in order to estimate sensor measurements and then compare them with the real ones. The result of this comparison is what makes pose correction possible. Remember that the difference between real measurements z_t and virtual measurements $h(\overline{\mu}_t)$ is required for correcting pose mean in EKF algorithm, as stated in Table 1.

3 Path Planning: "Base Points"

An original methodology is developed for path planning so that it benefits state estimation and, at the same time, maintains a good computational efficiency.

Base Points is an own path planning methodology developed upon three criteria: relative pose to target, change in direction and relative pose to the nearest obstacle. The definition of path would be through the computation of a series of points (base points) which will be specified by position coordinates in the map. Points will be generated with the same distance between one and the next. Therefore, it is only necessary to compute the angles that would relate the next position with the previous one. The following equation presents the three criteria in three terms, respectively.

$$\theta_i = \theta_{tar} - k_1 \Delta\theta_{tar} + k_2 \theta_{obst} \tag{6}$$

where:

- θ_i is the global reference angle between new and actual base point (radians)
- θ_{tar} is the global reference angle between target point and actual base point (radians)
- $\Delta\theta_{tar}$ is the difference between θ_{tar} and actual base point angle θ_{i-1}
- θ_{obst} is the difference between the global reference angle defined by the closest collision point (to obstacle) and actual base point, minus actual base point angle θ_{i-1}

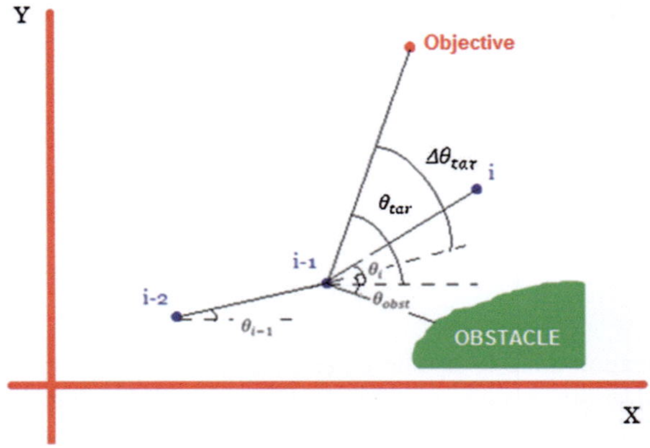

Fig. 3 Base points angle diagram

- k_1, k_2 are behavior coefficients that will be explained below (Fig. 3).

The *target related behavior coefficient* k_1 regulates how much the robot may change its direction in order to head for target point. If $\Delta\theta_{tar}$ has a higher value, it means that robot still cannot aim to target because it is avoiding some obstacle. On the contrary, if this angle is lower, it means robot is heading to target so there is no need for major correction as k_1 would be near to zero.

$$k_1 = 1 - e^{-k_d|\Delta\theta_{tar}|} \tag{7}$$

where k_d is a coefficient that programmer will assign by choice in order to regulate the value of k_1, and second term correction as well.

The *obstacle related behavior coefficient* k_2 will set an attraction or repulsion behaviors towards the nearest obstacle in the map. In probabilistic robotics, it is important to be near to obstacles in order to get the most information possible from environment because it will improve position tracking. Very close distance could be dangerous as it increases risk of collision, though. In addition, the more the robot approaches to its target, the less it would care about obstacles in the map. This will allow the robot to focus on target once it is close to it.

$$k_2 = \left(a_0 d_{obst} - \frac{a_1}{d_{obst}}\right)\left(\frac{d_{tar}}{d_{max}}\right)^n \tag{8}$$

where:

- d_{obst} is the distance to the closest collision point (obstacle)
- d_{tar} is the distance between target and actual base point
- d_{max} is the distance between the two furthest points among the map obstacles plus 2.2 times the distance between robot wheels.

- a_0, a_1 are coefficients used to tune up the threshold for attraction/repulsion behavior
- n is an index that programmer may assign by choice in order to regulate robot behavior towards obstacles. The higher the value, the less attention robot would pay to obstacles.

4 Results

Results of the application of position tracking and Base Points method for path planning on a CoroBot mobile are presented below.

For implementation, the variables shown on previous equations had the following values:

- *For odometric model*

  Wheel radius: $r = 5.22\, cm$
  Wheel distance: $b = 23.3\, cm$
  Motion covariance: $\Sigma_{\Delta} = \begin{bmatrix} 0.00303404|\Delta s_1| & 0 \\ 0 & 0.00324008|\Delta s_2| \end{bmatrix}$

- *For path planning*

  k_1 parameter: $k_d = 0.5$
  k_2 parameters: $a_0 = 0.0114$; $a_1 = 13.96$; $n = 0.5$

The algorithms explained during this paper were programmed on CoroBot using C language. In Fig. 4, it can be seen the results of the navigation system. Robot was asked to move from initial to target poses, avoiding two squared obstacles. Red points are the base points from path planning algorithm. Blue points are each position computed from EKF position tracking. Green points are theoretic sensor readings

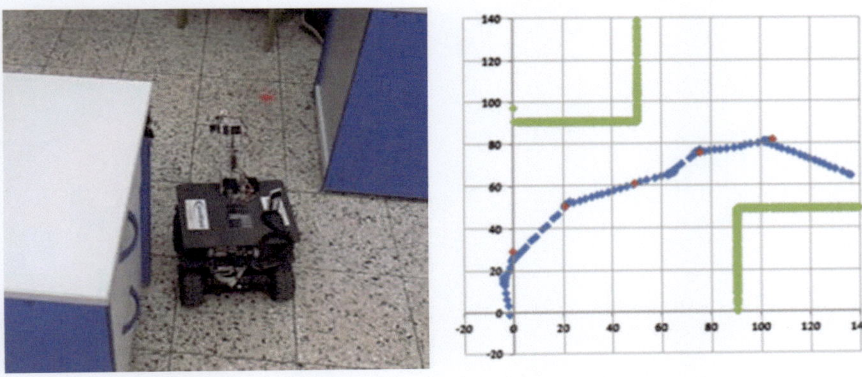

Fig. 4 Position tracking and path planning tested on CoroBot mobile

from map obstacles. Since the CoroBot unit used for this work did not have enough sensors for proper environment data feedback, real measurem ents were simulated by theoretic ones plus a white noise.

As the previous Figure shows, robot is capable of autonomously planning its own path in order to travel from initial to target pose. Furthermore, it is capable of following that path by tracking its own pose, which is computed on each iteration of EKF algorithm. The results prove that the unimodal Gaussian assumption for robot and environment states is valid for these cases in which map is static and initial pose is known beforehand.

References

Kozlowski K, Pazderski D (2004) Modeling and control of a 4-wheel skid-steering mobile robot, pp 1–20. Institute of Control and Systems Engineering, Poznan University of Technology, Poznan

Siegwart R, Nourbakhsh I (2004) Introduction to autonomous mobile robots, pp 160–190. The MIT Press, Massachusetts

Thrun S (2002) Is robotics going statistics? The field of probabilistic robotics, pp 2–5. School of Computer Science, Carnegie Mellon University, Pennsylvania

Suliman C, Cruceru C, Mondolveanu F (2009) Mobile robot position estimation using the Kalman filter, pp 1–3. Department of Automation, Transilvania University of Brasov, Brasov

Thrun S, Burgard W, Fox D (2006) Probabilistic robotics, pp 54–59. The MIT Press, Massachusetts

Kinematic Analysis of a Meso-Scale Parallel Robot for Laser Phonomicrosurgery

Sergio Lescano, Dimiter Zlatanov, Micky Rakotondrabe and Nicolas Andreff

Abstract The paper presents the kinematic model of a new meso-scale ($\sim$1 cm^3) parallel kinematic machine intended for laser phono-microsurgery of the vocal folds. The proposed mechanism uses the displacement generated by piezoelectric cantilever actuators manufactured via a Smart Composite Microstructure technique. The architecture, the geometry, and the position kinematics of the device, modeled as a spatial linkage, are discussed briefly. Then, the paper presents a velocity and singularity analysis and concludes that the new meso-scale parallel kinematic machine does not have singularities in the neighborhood of the reference configuration where it is required to operate.

Keywords Kinematic · Parallel kinematic machine · Screw theory · Piezoelectric · Phonosurgery

1 Introduction

Laser phonosurgery is a surgical procedure used to enhance the voice using a laser beam onto the vocal fold. This paper describes a meso-scale parallel kinematic machine (PKM) conceived with the aim to improve the quality of laser phono-surgery.

S. Lescano (✉) · M. Rakotondrabe · N. Andreff
Femto-ST Institute, Besançon, France
e-mail: sergio.lescano@femto-st.fr

M. Rakotondrabe
e-mail: mrakoton@femto-st.fr

N. Andreff
e-mail: nicolas.andreff@femto-st.fr

D. Zlatanov
University of Genoa, PMARlab Robotics Group, Genoa, Italy
e-mail: zlatanov@dimec.unige.it

A. Kecskeméthy and F. Geu Flores (eds.), *Interdisciplinary Applications of Kinematics*, Mechanisms and Machine Science 26, DOI 10.1007/978-3-319-10723-3_14

This mechanism is placed at the end of a laryngoscope and orients a mirror used to direct the laser beam onto the vocal fold. In order to account for all scale effects, to satisfy the constraints of the microworld, and to ensure the required accuracy, *piezo-electric material* (Ballas 2007; Clévy et al. 2011) combined with a *smart composite microstructure* (SCM) fabrication process (Wood et al. 2003, 2008) are proposed to build this meso scale PKM.

In Sect. 2, the kinematic requirements of the meso-structure are described, the mechanism's architecture and geometry are proposed, and the position kinematics is discussed in brief. Section 3 is devoted to velocity and singularity analysis using screw theory. The main conclusions are drawn and possibilities for future work are outlined in Sect. 4.

2 Kinematic Requirements and Proposed Structure

2.1 Kinematic Requirements

The PKM is located at the head of an adjustable laryngoscope. The laryngoscope's diameter is 20 mm, and the space assigned for the robot is $10 \times 10 \times 10\ mm^3$. The end of the laryngoscope is introduced into the cavity of the human larynx, at a distance of up to 20 mm from the vocal fold, as depicted in Fig. 1. To guide the direction of the laser beam onto the vocal fold, the PKM needs to have two rotational degrees of freedom (2 DOF). The standard average length of an adult vocal fold is between 17 and 20 mm. With a working distance of up to 20 mm, the PKM can scan the entire vocal fold if it can control the laser direction in a cone with a half-angle of 15°. The SCM manufacturing technique mentioned above, allows us to build the meso-scale robots as was shown by (Wood et al. 2003, 2008), first in a plane and then folding and popping out to form the 3D structure.

Minimal bandwidth of 200 Hz and a scanning resolution better than 100 μm are required to prevent overheating and damaging healthy tissue. Despite their small

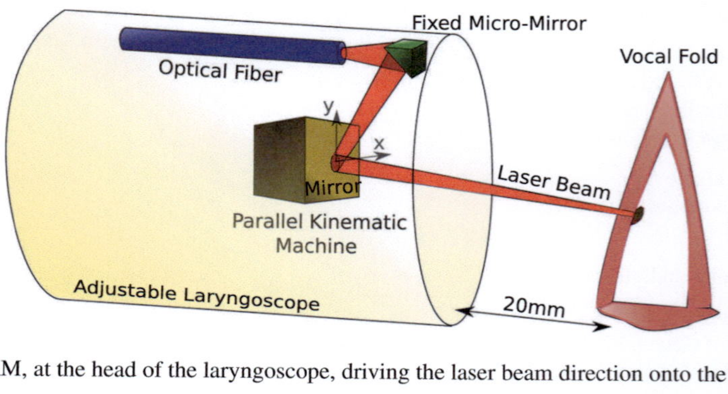

Fig. 1 PKM, at the head of the laryngoscope, driving the laser beam direction onto the vocal fold

displacement, piezoelectric cantilever actuators have been chosen for their nano-resolution, work speed, and adaptability to the required scale (Ballas 2007; Clévy et al. 2011). Piezoelectric cantilever actuators with dimensions of 10 mm × 2 mm reach displacements of 100 μm. Piezoelectric cantilevers are arranged in parallel to not surpass the assigned dimensions.

2.2 Proposed Mechanism Architecture and Geometry

The proposed PKM structure is depicted in Fig. 2. The end-effector, or the moving platform, is used to orient a mirror, Fig. 1.

The platform is connected to the base (attached to the laryngoscope) by a passive U-joint at P (leg 3), which constrains it to only tilt and pan. We define the base reference frame $Pxyz$ with coordinate unit vectors **i**, **j**, **k**, and a rotating frame, fixed to the platform, with origin at P and coordinate unit vectors **u**, **v**, **w**, Fig. 2.

The 2-dof platform motion is actuated with two identical 6-dof RUS legs, legs 1 and 2. Piezoelectric actuation is used and modeled as the first R-joint of the each actuated leg, i.e., the piezoelectric cantilever assumed to be a rigid rod rotating about

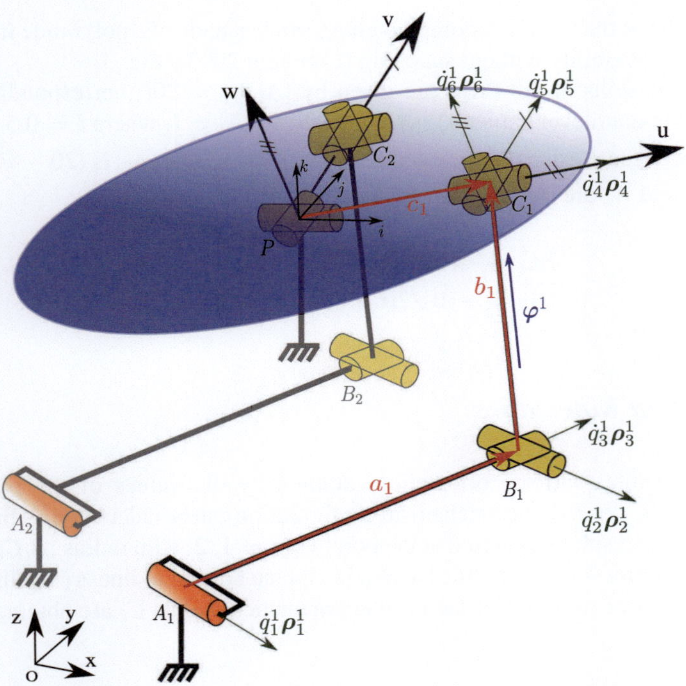

Fig. 2 Platform operated in parallel by two active legs and constrained by an unactuated U-joint leg

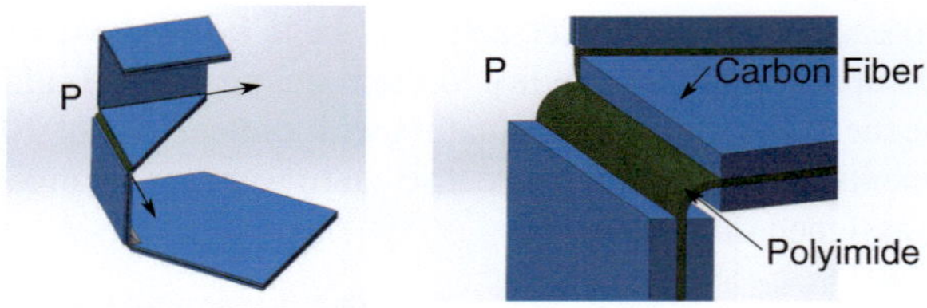

Fig. 3 Realization of the passive U-joint

a fixed axis. The two active joints have a common axis, A_2A_1, with unit vector $\mathbf{k}_1 = \sqrt{2}/2\mathbf{i} - \sqrt{2}/2\mathbf{j}$. (Point A_i is the projection of the U-joint center B_i on this axis.)

In each active leg, the axis of joint 2, the first joint of the passive U-joint centered at $B_i, i = 1, 2$, is parallel to A_1A_2, while the center, C_i, of the spherical joint is along the common perpendicular of the U-joint axes.

At the reference configuration, when the angles of the U-joint at P are $\theta_1 = \theta_2 = 0$, we have $\mathbf{c}_1 = \overrightarrow{PC_1}$ parallel to $\mathbf{u} = \mathbf{i}$ and $\mathbf{c}_2 = \overrightarrow{PC_2}$ parallel to $\mathbf{v} = \mathbf{j}$. Moreover, at the zero configuration, the distal links, B_iC_i, are parallel to the z axis, while the proximal links, A_iB_i, are perpendicular to it and at equal angles with the x and y axes.

Each joint of the PKM is a compliant hinge and is made of a polyimide film which allows great flexibility without snapping (Lobontiu 2003), Fig. 3.

The dimensions of the PKM are given by $|A_iB_i| = 20l$ (corresponding to the piezoelectric-cantilever actuator), $|B_iC_i| = 2l$, $|PC_i| = l$, where $l = 0.5$ mm. The distance between Pz and the common actuated-joint axis, A_1A_2, is $(20-\frac{\sqrt{2}}{2})l$, while points A_1 and A_2 are at $\sqrt{2}l$ from each other. Therefore,

$$\begin{aligned} \overrightarrow{PA_1} &= -(10\sqrt{2}-1)l\mathbf{i} - 10\sqrt{2}l\mathbf{j} - 2l\mathbf{k} \\ \overrightarrow{PA_2} &= -10\sqrt{2}l\mathbf{i} - (10\sqrt{2}-1)l\mathbf{j} - 2l\mathbf{k} \end{aligned} \tag{1}$$

2.3 Position Kinematics

Given a feasible platform orientation, defined by the values of the passive-leg U-joint angles, (θ_1, θ_2), the mechanism configuration can be calculated by finding the intersection of a sphere centered at the point $C_i, i = 1, 2$, with radius $|B_iC_i|$ and the circle with center A_i, with radius $|A_iB_i|$, in a plane normal to line A_1A_2. In general, there will be four solutions of the inverse kinematics. Points C_i are obtained as:

$$\overrightarrow{PC_1} = l\mathbf{u} = c_2 l\mathbf{i} + s_1 s_2 l\mathbf{j} - c_1 s_2 l\mathbf{k} \tag{2}$$

$$\overrightarrow{PC_2} = l\mathbf{v} = c_1 l\mathbf{j} + s_1 l\mathbf{k} \tag{3}$$

where $c_1 = \cos\theta_1$, $s_1 = \sin\theta_1$. Indeed, the platform orientation is given by

$$R(\theta_1, \theta_2) = R_x(\theta_1) R_y(\theta_2) = \begin{bmatrix} c_2 & 0 & s_2 \\ s_1 s_2 & c_1 & -s_1 s_2 \\ -c_1 s_2 & s_1 & c_1 c_2 \end{bmatrix} \tag{4}$$

We have $\overrightarrow{B_j C_j} = \overrightarrow{A_j C_j} - \overrightarrow{A_j B_j}$, and so

$$|\overrightarrow{B_j C_j}|^2 = 4l^2 = |\overrightarrow{A_j C_j}|^2 + 400l^2 - 2\overrightarrow{A_j C_j} \cdot \overrightarrow{A_j B_j} \tag{5}$$

where $\overrightarrow{A_j B_j} = 10l(\sqrt{2}c_1^j \mathbf{i} + \sqrt{2}c_1^j \mathbf{j} + 2s_1^j \mathbf{k})$, and $c_i^j = \cos q_i^j$, $s_i^j = \sin q_i^j$. Using (1)–(3) and (5) it is now possible to obtain a linear equation in terms of c_1^j, s_1^j

$$F_j c_1^j + G_j s_1^j = H_j \tag{6}$$

where F_j, G_j, and H_j depend only on θ_1, θ_2, resulting in two solutions for q_1^1.

E.g. for $\theta_1 = 15°$, $\theta_2 = -15°$ the solutions for (q_1^1, q_1^2) are:

$$\begin{aligned} &(0.720102952001288°, 0.742258398498497°) \\ &(0.720102952001288°, 12.160623836841685°) \\ &(12.16311819616120381°, 0.742258398498497°) \\ &(12.16311819616120381°, 12.160623836841685°) \end{aligned} \tag{7}$$

3 Velocity and Singularity Analysis

The velocity analysis is performed using screw theory, which has found a wide application, in particular in the study of parallel robots. Instantaneous motion and forces acting on a rigid body can be represented by a twist and a wrench, respectively. With respect to a given reference frame, a twist (or a wrench) is given by a pair of (non-commensurate) physical vectors, $(\boldsymbol{\omega}, \mathbf{v})$, the body's angular velocity and the linear velocity at the origin (or $(\mathbf{f}, \mathbf{m})$, the resultant force and moment with respect to the origin). A normalized twists or wrenches has the same 3D geometrical representation, called a screw, defined by an axis and a pitch. (For a translation or a couple, there is no axis, but only a direction, and the pitch is said to be infinite.)

The reciprocal product, $\boldsymbol{\omega} \cdot \mathbf{m} + \mathbf{v} \cdot \mathbf{f}$, measures the power exerted by a wrench on a twist. Two screws, and any wrench-twist pair on them, are said to be reciprocal to each other if the product is zero.

3.1 Velocity Kinematics

On a PKM the instantaneous twist, $\boldsymbol{\xi}$, of the platform with respect to the ground can be expressed as a linear combination of the joint screws of leg j (Mohamed and Duffy 1985). Denoting by $\dot{q}_i^j$ and $\boldsymbol{\rho}_i^j$ the amplitude and the unit zero-pitch twist joint twist associated with the ith joint of the jth leg, we have:

$$\boldsymbol{\xi} = \dot{q}_1^1 \boldsymbol{\rho}_1^1 + \dot{q}_2^1 \boldsymbol{\rho}_2^1 + \dot{q}_3^1 \boldsymbol{\rho}_3^1 + \dot{q}_4^1 \boldsymbol{\rho}_4^1 + \dot{q}_5^1 \boldsymbol{\rho}_5^1 + \dot{q}_6^1 \boldsymbol{\rho}_6^1 \tag{8a}$$

$$\boldsymbol{\xi} = \dot{q}_1^2 \boldsymbol{\rho}_1^2 + \dot{q}_2^2 \boldsymbol{\rho}_2^2 + \dot{q}_3^2 \boldsymbol{\rho}_3^2 + \dot{q}_4^2 \boldsymbol{\rho}_4^2 + \dot{q}_5^2 \boldsymbol{\rho}_5^2 + \dot{q}_6^2 \boldsymbol{\rho}_6^2 \tag{8b}$$

$$\boldsymbol{\xi} = \dot{\theta}_1 \boldsymbol{\rho}_1^3 + \dot{\theta}_2 \boldsymbol{\rho}_2^3 \tag{8c}$$

Since legs 1 and 2 have 6 dof they impose no constraint on the platform. Therefore the possible twists, $\boldsymbol{\xi}$, at every configuration, are given by (8c) as any linear combination of the passive leg's two revolute-joint unit twists, $\boldsymbol{\rho}_1^3$ and $\boldsymbol{\rho}_2^3$. When such a well-defined continuously-changing basis exists for the twist system of the platform, the velocity kinematics can be expressed with as few equations as the parallel mechanism's dof (Zoppi et al. 2006).

For this, one can first substitute (8c) into (8a) and (8b), and then eliminate the passive joint velocities using reciprocal screws obtaining a system of two equations relating the platform angular speeds $\dot{\theta}_1$ and $\dot{\theta}_2$ with the input velocities $\dot{q}_1^1$ and $\dot{q}_1^2$.

For each active leg, there is a unit pure force, $\boldsymbol{\varphi}^j$, $j = 1, 2$, reciprocal to all five passive-joint screws. Its axis passes through the centers of the leg's U- and S-joints.

(In a leg-singular configuration, where the center of the spherical joint is in the plane of the U-joint, there will be a whole pencil of concurrent planar forces with this property, and two, rather than one, reciprocal screws for that leg must be used to obtain the input-output velocity equations. We will assume that such leg postures do not occur during the operation of the mechanism.)

Taking the reciprocal product of $\boldsymbol{\varphi}^1$ with (8a) and of $\boldsymbol{\varphi}^2$ with (8b), we obtain:

$$\begin{bmatrix} \boldsymbol{\varphi}^1 \cdot \boldsymbol{\rho}_1^3 & \boldsymbol{\varphi}^1 \cdot \boldsymbol{\rho}_2^3 \\ \boldsymbol{\varphi}^2 \cdot \boldsymbol{\rho}_1^3 & 0 \end{bmatrix} \begin{bmatrix} \dot{\theta}_1 \\ \dot{\theta}_2 \end{bmatrix} = \begin{bmatrix} \boldsymbol{\varphi}^1 \cdot \boldsymbol{\rho}_1^1 & 0 \\ 0 & \boldsymbol{\varphi}^2 \cdot \boldsymbol{\rho}_2^3 \end{bmatrix} \begin{bmatrix} \dot{q}_1^1 \\ \dot{q}_1^2 \end{bmatrix} \tag{9}$$

In (9), the term $\boldsymbol{\varphi}^2 \cdot \boldsymbol{\rho}_2^3$ is zero, because the joint-2 axis of the passive leg always passes through point C_2. To calculate the other entries, we can use:

$$\boldsymbol{\rho}_1^3 = \begin{bmatrix} \mathbf{i} \\ 0 \end{bmatrix}, \quad \boldsymbol{\rho}_2^3 = \begin{bmatrix} \mathbf{v} \\ 0 \end{bmatrix}, \quad \boldsymbol{\varphi}^j = \begin{bmatrix} \frac{\mathbf{b}^j}{|\mathbf{b}^j|} \\ \mathbf{c}_j \times \frac{\mathbf{b}^j}{|\mathbf{b}^j|} \end{bmatrix}, \quad \boldsymbol{\rho}_1^j = \begin{bmatrix} \mathbf{k}_1 \\ \overrightarrow{PA_1} \times \mathbf{k}_1 \end{bmatrix}, \quad j = 1, 2 \tag{10}$$

I Since $\overrightarrow{PA_j} = \mathbf{c}^j - \mathbf{b}^j - \mathbf{a}^j$ we have that

$$\begin{aligned}\boldsymbol{\varphi}^j \cdot \boldsymbol{\rho}_1^j &= \frac{\mathbf{b}^j}{|\mathbf{b}^j|} \cdot (\mathbf{c}^j - \mathbf{b}^j - \mathbf{a}^j) \times \mathbf{k}_1^j \; + \; \mathbf{k}_1^j \cdot (\mathbf{c}^j \times \frac{\mathbf{b}^j}{|\mathbf{b}^j|}) \\ &= \frac{\mathbf{b}^j}{|\mathbf{b}^j|} \cdot [\mathbf{k}_1^j \times \mathbf{a}^j] \qquad\qquad\qquad j = 1, 2\end{aligned} \tag{11}$$

and so (9) can be written as:

$$\begin{bmatrix} \frac{\mathbf{b}^1}{|\mathbf{b}^1|} \cdot \mathbf{k}_1 \times \mathbf{a}^1 & 0 \\ 0 & \frac{\mathbf{b}^2}{|\mathbf{b}^2|} \cdot \mathbf{k}_1 \times \mathbf{a}^2 \end{bmatrix} \begin{bmatrix} \dot{q}_1^1 \\ \dot{q}_1^2 \end{bmatrix} = \begin{bmatrix} \mathbf{c}^1 \times \frac{\mathbf{b}^1}{|\mathbf{b}^1|} \cdot \mathbf{i} & \mathbf{c}^1 \times \frac{\mathbf{b}^1}{|\mathbf{b}^1|} \cdot \mathbf{v} \\ \mathbf{c}^2 \times \frac{\mathbf{b}^2}{|\mathbf{b}^2|} \cdot \mathbf{i} & 0 \end{bmatrix} \begin{bmatrix} \dot{\theta}_1 \\ \dot{\theta}_2 \end{bmatrix} \tag{12}$$

which can be abbreviated as $B\dot{\mathbf{q}} = D\dot{\boldsymbol{\theta}}$, where B is a diagonal matrix.

3.2 Singularity Analysis

Gosselin and Angeles suggested a classification of singularities for parallel manipulators into three main groups associated with the singularity of square matrices such as D and B above (Gosselin and Angeles 1990). Later Zlatanov gave a more comprehensive study of the singularities of closed-loop mechanisms including configurations where the matrices D and B are not square (Zlatanov et al. 1994, 1998). This can occur in the case mentioned above of either leg 1 or leg 2 being in a singular posture with the center of the S-joint lying in the plane of the U-joint, in which case a redundant passive mobility singularity is present. However, if we assume that such leg postures are outside of the range of operation of the mechanism, the singularities of the mechanism are the configurations where either B or D is singular, leading to a degeneration of the inverse or the direct instantaneous kinematics, respectively.

In (9) or (12), matrix B is singular when one or both elements of its diagonal becomes zero, signifying that the actuated joint axis intersects one of the reciprocal force axes. In this case, if D is nonsingular, we have a singularity of type (IO, RI), where the end-effector loses freedoms and the input velocities can be nonzero with a stationary platform.

This cannot happen in a neighborhood of the reference configuration where $\mathbf{q}_1^j = 0$, matrix B is equal to diag($|\mathbf{a}_1^1|$, $|\mathbf{a}_2^1|$) far from having a zero determinant.

To find a singularity of the matrix D, the axis of either $\boldsymbol{\varphi}^1$ or $\boldsymbol{\varphi}^2$ must lie in the plane of the passive-leg U-joint. In this case, the platforms freedoms cannot be controlled. It will be able to move with actuators locked and the actuated velocities are not independent and cannot be chosen arbitrarily, a singularity of types RO and II. Here again, with a good choice of geometric parameters, the PKM is safe in a neighborhood of the position where $\mathbf{q}_1^j = 0$ with

$$D = \begin{bmatrix} 0 & -|\mathbf{c}^1| \\ |\mathbf{c}^2| & 0 \end{bmatrix} \tag{13}$$

Since the PKM workspace is limited to the angles required to scan entire vocal fold ($\pm 15°$), both B and D are expected nonsingular, in this reduced workspace.

4 Conclusion

This paper presents a meso-scale PKM dedicated to the laser phonomicrosurgery of vocal folds. A kinematic analysis is performed. Reciprocal screws are used to find the two matrices that describe the mechanism's velocity behavior. All possible cases of singularities are analyzed. The limitation of the workspace and the proposed mechanism geometry avoid each type of singularity.

In the future, owing to the microfabrication limits and the robot structure compliance, a study of the directional variation of the screws will be taken into account using interval analysis (Merlet 2009). Moreover, due to the low displacement generated by piezoelectric cantilever actuators, a different configuration in a position near-singular for the matrix D could be exploited (Voglewede and Ebert-Uphoff 2002) to generate large platform rotations with small displacements of the actuators, being near the point of singularity without reaching it to preserve controllability.

Acknowledgments This work has been supported by the European Union Seventh Framework Programme [FP7-ICT-2011-7]. Project μRALP (www.microralp.eu).

References

Ballas R (2007) Piezoelectric multilayer beam bending actuators: static and dynamic behavior and aspects of sensor integration. Springer, Microtechnology and Mems

Clévy C, Rakotondrabe M, Chaillet N (2011) Signal measurement and estimation techniques for micro and nanotechnology. Springer, New York

Gosselin C, Angeles J (1990) Singularity analysis of closed-loop kinematic chains. IEEE Trans Robot Autom 6(3):281–290

Lobontiu N (2003) Compliant mechanisms: design of flexure hinges. CRC Press, Mechanical engineering

Merlet JP (2009) Interval analysis and reliability in robotics. Int J Reliab Safe 3

Mohamed MG, Duffy J (1985) A direct determination of the instantaneous kinematics of fully parallel robot manipulators. J Mech Trans Autom Des 107

Voglewede PA, Ebert-Uphoff I (2002) Two viewpoints on the unconstrained motion of parallel manipulators at or near singular configurations. In: International Conference on Robotics and Automation, pp 503–510

Wood R, Avadhanula S, Menon M, Fearing R (2003) Microrobotics using composite materials: the micromechanical flying insect thorax. IEEE Int Conf Robot Autom 2:1842–1849

Wood RJ, Avadhanula S, Sahai R, Steltz E, Fearing RS (2008) Microrobot design using fiber reinforced composites. J Mech Des 130(5):052,304+

Zlatanov D, Benhabib B, Fenton R (1994) Analysis of the instantaneous kinematics and singular configurations of hybrid-chain manipulators. Proceedings of the ASME 23rd Biennial Mechanisms Conference 72:467–476 (1994)

Zlatanov D, Fenton R, Benhabib B (1998) Identification and classification of the singular configurations of mechanisms. Mech Mach Theory 33(6):743–760

Zoppi M, Zlatanov D, Molfino R (2006) On the velocity analysis of interconnected chains mechanisms. Mech Mach Theory 41(11):1346–1358

Characteristics of a Walking Simulator with Parallel Manipulators

Dante Elías, Rocío Callupe and Marco Ceccarelli

Abstract In this paper an on-site walking simulator with parallel manipulators is presented by looking at characteristics of a prototype built in Lima, Perú. Design peculiarities of parallel architectures are used for a powerful precise walking simulator by combining Kinematic features with dynamic capabilities through a suitable control system.

Keywords Gough-Stewart platform · Walking simulator · Parallel manipulator

1 Introduction

Human walking addresses great attention both as nature inspiration for biped and humanoid robots and as matter of study for biomedical applications. Thus, the study of human walking is worked out from engineering viewpoints in theoretical approaches to propose biomechanical models (Levine et al. 2012), and in practical designs to develop walking simulators.

In particular walking simulators attract great interests for applications in sport training and physiotherapy exercising (Deutsch et al. 2008). Examples of solutions can be pointed out in increasing the technical capacity of high-performance athletes within gymnasium sport applications and in people with walking problems caused by a neurological disorder within medical environments. Main challenges of existing solutions can be recognized in their limited capabilities in actions and displacements

D. Elías (✉) · R. Callupe
Departamento de Ingeniería, Laboratorio de Investigación en Biomecánica y Robótica Aplicada, Pontificia Universidad Católica del Perú, Lima, Perú
e-mail: delias@pucp.edu.pe

R. Callupe
e-mail: rcallup@pucp.edu.pe

M. Ceccarelli
Laboratory of Robotics and Mechatronics, DiCeM—University of Cassino and South Latium, Cassino, Italy
e-mail: ceccarelli@unicas.it

A. Kecskeméthy and F. Geu Flores (eds.), *Interdisciplinary Applications of Kinematics*, Mechanisms and Machine Science 26, DOI 10.1007/978-3-319-10723-3_15

for long duration walking. The on-site solutions are based on treadmills that do not give the possibility of a practical walking assistance whereas the walking simulators that are based on exoskeletons require power units on board with fairly heavy solution (Yoon et al. 2010).

In this paper a walking simulator combining peculiarities of on-site motion and powerful guiding is presented with a design that is based on the operation of two parallel manipulators (Sevillano et al. 2011).

The design characteristics and operation features are presented to characterise the Lima Walking Simulator whose functioning can be used in a variety of modes with an on-site walking of an operator/patient. A built prototype has been used in tests both to validate the designed system and to explore performance capabilities in applications for diagnosis and rehabilitation of human walking as depending of the level of motion/force guiding.

2 Characteristics for Walking Simulator

Human walking is characterized as bipedal synchronized motion by referring to characteristic planes in sagittal and front directions, Fig. 1. Typical values are listed in Table 1 as data for operation and design feature of a walking simulator.

The kinematics of human walking can be summarized in the leg phases for support and swing with proper motion of laws in the articulations as well as in the leg coordinated motion (Ceccarelli 2004; Hreljac et al. 2007; Levine et al. 2012). But

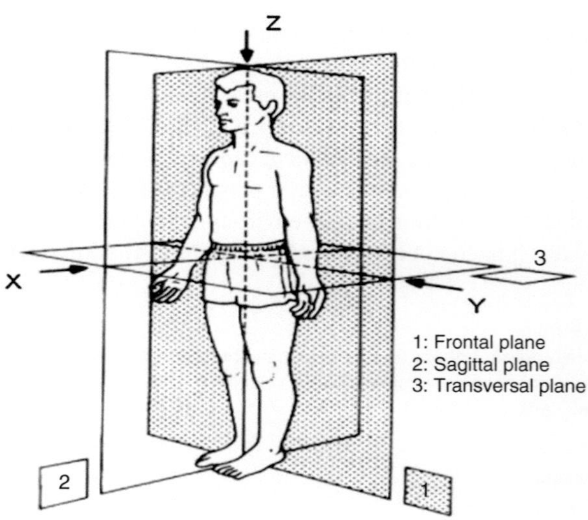

Fig. 1 Reference axes and characteristic planes for human walking

Table 1 Numerical values for a walking simulator with reference as in Fig. 1

Variable of interest	Numerical values
Maximum static load, Z axes	700 N
Maximum static load, X axes	200 N
Maximum static load, Y axes	100 N
Maximum displacement, Z axes	0–150 mm
Maximum displacement, X axes	± 400 mm
Maximum displacement, Y axes	± 50 mm
Maximum rotation, axes Z	±5°
Maximum rotation, axes X	±5°
Maximum rotation, axes Y	±30°
Maximum linear velocity, X axes	1,834 m/s
Maximum angular velocity, Y axes	± 3.5 rad/s

specific values can be identified for each patient in rehabilitation training and along its evolution.

Dynamic properties of human walking are related to payload characteristics due to muscular power of legs. General data can be considered in the characteristic planes in Fig. 1 as summarized in Table 1.

In general, a human leg is considered with 3 dofs and a foot is modelled with 2 dofs to achieve a normal capability of 1,200 N including body weight in a regular walking of 1 step/s, when the total body weights 700 N.

A walking simulator is asked to assist and/or train patients in walking exercises with the aim to perform a normal walking or specific motions with exercise characteristics within suitable ranges towards a normal walking (Riener et al. 2006; Yoon et al. 2010). The characteristics of normal human walking as summarized above in Table 1, Fig. 1, are reference values for requirements and constraints in design and operation of walking simulators that nevertheless should be flexible to fairly easily adapt to the specific conditions and physiotherapy of a patient mainly in term of step size, step speed, walking condition (normal, sloped plane), motion articulations, ground impact of feet, and possibility of nursery assistance and help.

3 Design and Model of Lima Walking Simulator

Since a main focus can be addressed to combining step characteristics with load capabilities, in Lima a novel structure for walking simulator has been conceived for an on-site walking exercise by using a couple of movable platforms for foot allocations (Anchante 2010; Boian et al. 2005; Sevillano et al. 2011). Movable platforms guide the feet on step motion with prescribed force. Thus, it has been thought convenient to use parallel manipulator structures whose mobile plate is the foot platform.

Each parallel manipulator has been designed to carry up to 800 N for step motion area with 80 cm of step length and 50 cm of step height.

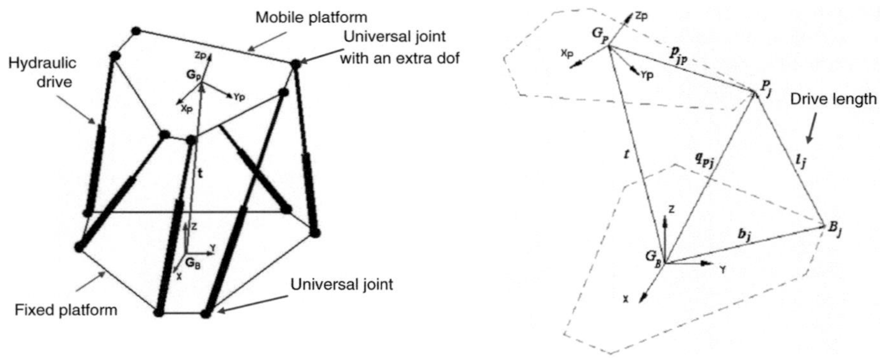

Fig. 2 Model of the parallel manipulators for the structure guiding the foot platforms

Table 2 Design sizes and operation characteristics of the parallel manipulators for foot platform

D (maximum parallel distance between mobile and fixed platforms)	700 mm
r_{mobile} (radius of mobile platform)	150 mm
r_{fixed} (radius of fixed platform)	215 mm
Stroke of linear drives	200 mm
Maximum drive force	3,300 N

The model of Fig. 2 has been used for the chosen structure of parallel manipulators in design calculations and motion programming.

In particular, with the above operation requirements the driving parallel manipulators have been designed with the characteristics in Table 2.

Simulations have been computed to check the feasibility of platform motions as well as to identify the operation features (Anchante 2010; Guo and Li 2006; Lopes 2009; Sevillano et al. 2011; Staicu 2011). In Fig. 3 actuation force of the drives are computed with a maximum value of 3,300 N.

In addition, Tables 3 and 4 reports computed numerical values for a design characterization of the structure and drives of the parallel manipulators for a 60 step/min of cadence with 8 points of reference.

As regarding with workspace characteristics (Castelli et al. 2008; Ceccarelli 2004; Jiang and Gosselin 2009; Merlet 2006), simulations have been carried out to check motion capabilities as in the examples of Fig. 4.

Figure 5 shows the structure of the control system that has been designed and implemented to program and regulate the Lima Walking Simulator during the operation by using a PC. Control system is based on algorithms for position control whereas the power is regulated in a unit of a hydraulic station for the hydraulic drivers in the platform (Davliakos and Papadopoulos 2008; Pi and Wang 2011). The adopted solution is based on market components of common industrial automation with hydraulic systems (Bances et al. 2012).

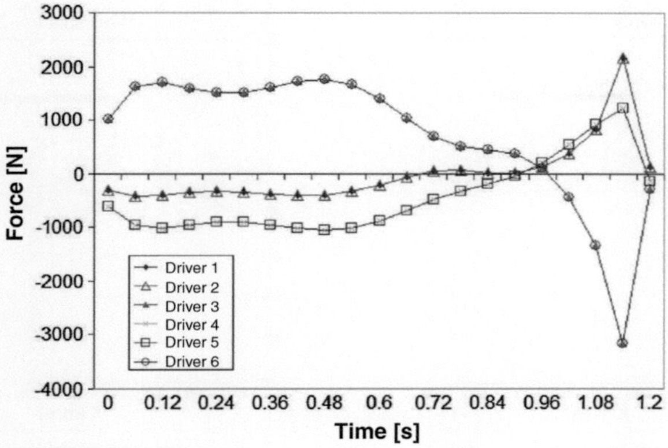

Fig. 3 Computed values for the drives of the parallel manipulator in Fig. 1 as foot platform

Table 3 Computed reaction forces as a function of pose of the foot platform as function of time

		Conditions in the center of gravity of the mobile platform								
		Position (mm)			Orientation (deg)			Reaction force (N)		
Point	t[s]	X	Y	Z	rot. Z = Ψ	rot. Y = θ	rot. X = Φ	X	Y	Z
0	0	−400	0	650	0	−28	0	−620	0	−700
1	0.0444	−300	0	700	0	−28	0	−570	0	−600
2	0.0889	−200	0	680	0	−30	0	−520	0	−500
3	0.1333	−100	0	650	0	−15	0	−470	0	−400
4	0.1778	0	0	620	0	−5	0	−420	0	−300
5	0.2222	100	0	600	0	0	0	470	0	−400
6	0.2667	200	0	580	0	0	0	520	0	−500
7	0.3111	300	0	600	0	5	0	570	0	−600
8	0.3556	400	0	650	0	28	0	620	0	−700

Table 4 Computed forces in the drives as function of time [N]

Point		0	1	2	3	4	5	6	7	8
	1	−1655.3	−1532	−537.7	−1163.1	948.75	1119.2	1465.8	2009.3	2201.6
	2	2201.6	1844.3	923.2	1162.5	−867.03	−1228.1	−1489.4	−1921.7	−1655.3
Drive	3	1028.9	636.1	372.2	164.5	−106.38	−298.2	−352.2	−557.7	−1770
	4	−3292.7	−2384.8	−805.1	−1102.5	791.85	882.6	1144	1654.6	3202
	5	3202	2305.7	1040.2	1009.7	−697.06	−1033.3	−1271.6	−1783	−3292.7
	6	−1770	−1206	−360.6	−400.8	233.87	133.8	190.6	305.6	1028.9

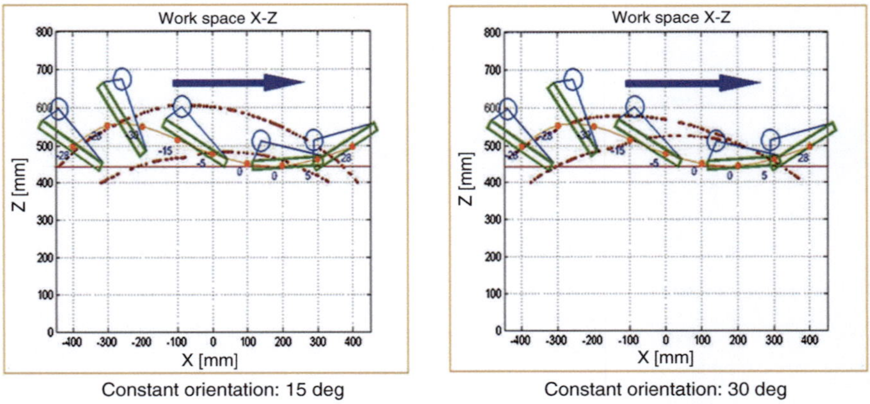

Fig. 4 Examples of computed workspace area for feasible motion of the foot platforms

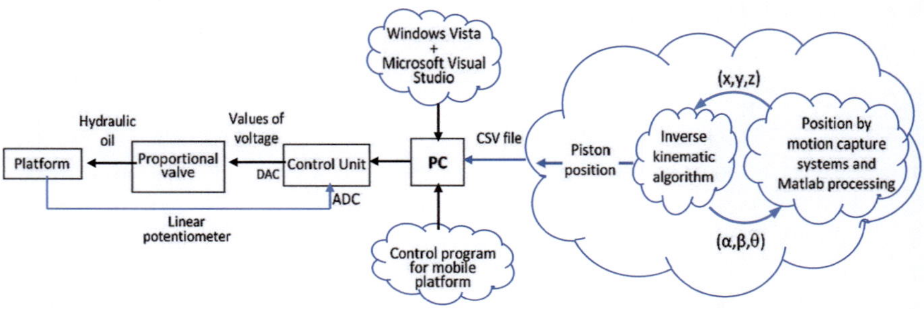

Fig. 5 Scheme of the control system for the Lima walking simulator

4 Experiences with a Prototype

A prototype has been built in the Laboratory of Research in Biomechanics and Applied Robotics at the Pontificia Universidad Católica del Perú in Lima with a first layout in 2008 (Sevillano et al. 2011). In Fig. 6 the prototype of Lima Walking Simulator is shown as built with a hydraulic actuation with 12 proportional hydraulic valves in a power unit of 3 kW.

On the mobile plates suitable shoes are installed as shown in Fig. 6b to accommodate the feet of a patient. Several tests have been worked out with a layout as in Fig. 7 where a patient walks on site with complementary equipment that permit him/her to stand vertically in comfortable position.

An on-site walking test is illustrated in the snapshots of Fig. 8 in which it is appreciable how the foot platforms guide the walking simulation by pushing on the feet to perform walking motion of the leg with proper articulation movements.

Output of a test is the walking motion with prescribed step size and by adjusting the pushing force through a regulation of the hydraulic power system. A test can be

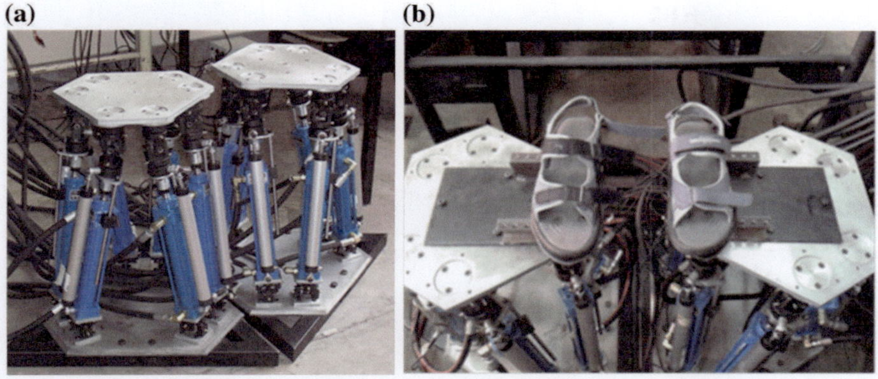

Fig. 6 A photo of the built prototype of Lima walking simulator in Lima: **a** the two parallel manipulators; **b** the foot platforms with shoes connectors

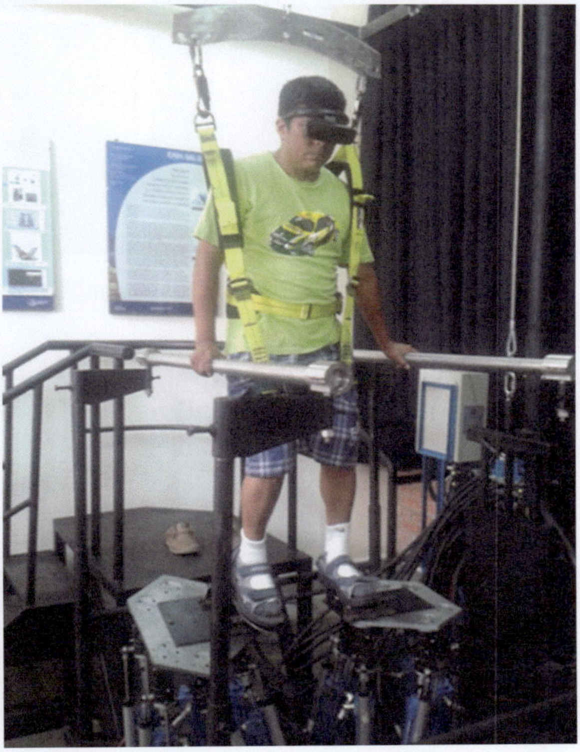

Fig. 7 A lab layout for tests with the prototype of Lima walking simulator

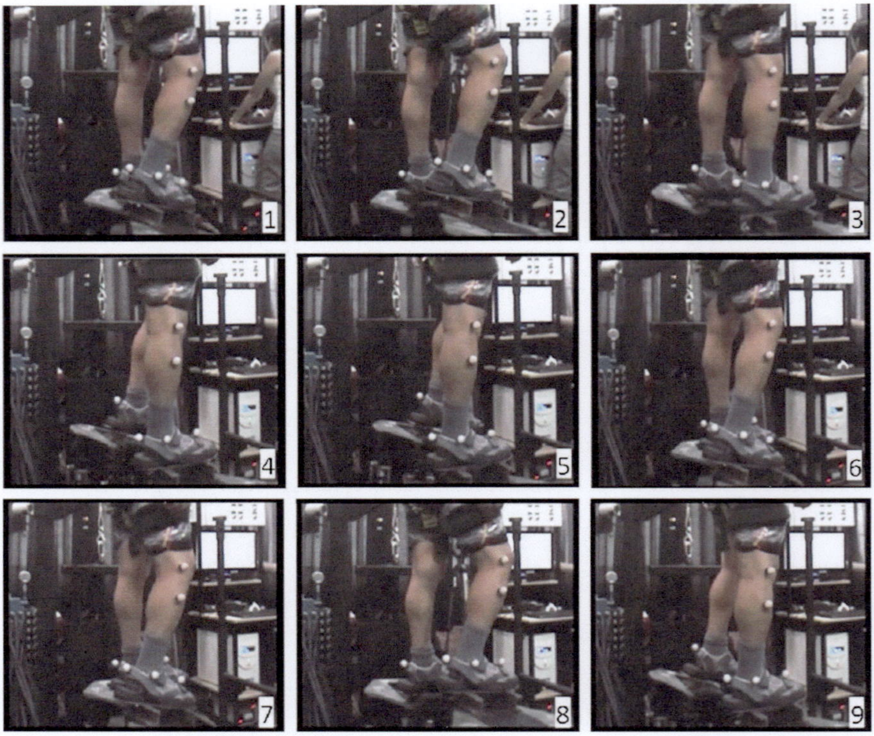

Fig. 8 Snapshot of an on-site walking test with Lima walking simulator

planned according to requirements of patient weight and prescribed action on the feet from a full guided walking up to slight assisted exercise.

5 Conclusions

Laboratory experiences with a built prototype in Lima have shown a successful application of a walking simulator as rehabilitation platform with a specific on-site operation mode. The prototype design has combined motion planning with dynamic performance in two parallel Gough-Stewart platforms to obtain an on-site walking simulator with high payload capacity and precise synchronized repeatable movements.

Acknowledgments The supports of governmental Presidency of Peru and of the Directorate of Research Management of Pontificia Universidad Católica del Perú are gratefully acknowledged through grant of the FINCyT and of the LUCET program, respectively. The collaboration among the authors has been supported by Pontificia Universidad Católica del Perú during a sabbatical visit of professor Ceccarelli in Lima in 2012.

References

Anchante C (2010) Modelación y simulación dinámica del mecanismo paralelo tipo plataforma Stewart-Gough para un simulador de marcha. Mechanical Engineering Thesis, PUCP

Bances E, Chang J, Elías D (2012) Sistema de control del simulador de marcha tipo pie plataforma de la PUCP. In: Sotomayor J, Pérez C, Rivas R (eds) XV Congreso Latinoamericano en Control Automático, pp. 128 (full text in CD). Pontificia Universidad Católica del Perú, Lima

Boian RF, Bouzit M, Burdea GC, Lewis J, Deutsch J (2005) Dual Stewart platform mobility simulator. IEEE Conference on Rehabilitation Robotics, pp 550–555

Castelli G, Ottaviano E, Ceccarelli MA (2008) Fairly general algorithm to evaluate workspace characteristics of serial and parallel manipulators. Int J Mech Based Des Struct Mach 36:14–33

Ceccarelli M (2004) Fundamentals of mechanics of robotic manipulation. Kluwer Academic Publishers/Springer, The Netherlands

Davliakos I, Papadopoulos E (2008) Model-based control of a 6-dof electrohydraulic Stewart-Gough platform. Mech Mach Theor 43:1385–1400

Deutsch JE, Boian RF, Lewis JA, Burdea GC, Minsky A (2008) Haptic effects modulate kinetics of gait but not experience of realism in a virtual reality walking simulator. IEE Conference on Virtual Rehabilitation, pp 36–40

Guo H, Li H (2006) Dynamic analysis and simulation of a six degree of freedom Stewart platform manipulator. Proceedings of the Institution of Mechanical Engineers. J Mech Eng Sci 220:61–72

Hreljac A, Imamura RT, Escamilla RF, Edwards WB (2007) When does a gait transition occur during human locomotion? J Sports Sci Med 6:36–43

Jiang Q, Gosselin CM (2009) Determination of the maximal singularity-free orientation workspace for the Gough-Stewart platform. Mech Mach Theor 44:1281–1293

Levine D, Richards J, Whittle MW (2012) Whittle's Gait anal, 5th edn. Churchill Livingstone, London

Lopes AM (2009) Dynamic modeling of a Stewart platform using the generalized momentum approach. Comm Nonlinear Sci Numer Simul 14:3389–3401

Merlet J (2006) Parallel robots, 2nd edn. Springer, The Netherlands

Pi Y, Wang X (2011) Trajectory tracking control of a 6-DOF hydraulic parallel robot manipulator with uncertain load disturbances. Control Eng Pract 19:185–193

Riener R, Lünenburger L, Colombo G (2006) Human-centered robotics applied to gait training and assessment. J Rehabil Res Dev 43:679–694

Sevillano G, Elias D, Callupe R, Marcacuzco R, Barriga B (2011) Gait simulator based on the parallel Stewart-Gough platform. In: Kecskemethy A, Potkonjak V, Müller A (eds) Interdisciplinary applications of kinematics, pp 99–108. Springer, The Netherlands

Staicu S (2011) Dynamics of the 6–6 Stewart parallel manipulator. Robot Comput-Int Manuf 27:212–220

Yoon J, Novandy B, Yoon ChH, Park KJ (2010) A 6-DOF Gait rehabilitation robot with upper and lower limb connections that allows walking velocity updates on various terrain. IEEE/ASME Trans Mech 15:201–215

Prototype Upper Limb Prosthetic Controlled by Myoelectric Signals Using a Digital Signal Processor Platform

Ulises Gordillo Zapana, Renée M. Condori Apaza, Nancy I. Orihuela Ordoñez and Alfredo Cárdenas Rivera

Abstract In this research project, we aimed to design and implement an upper limb prosthesis controlled by myoelectric signals using a digital signal processor platform. To emulate the seven main movements of a human arm, a robotic arm was produced that was capable of using the control signals generated by a human arm, where we captured the electrical pulses to design a silver/silver chloride contact type surface electrode using a plating process in a chemical laboratory. This method is an alternative technological support for amputees or partially paralyzed muscles, which typically remain intact so they can exercise control. The signals produced by these muscles can operate a prosthesis or a robotic device. Therefore, the prototype arm design process comprised the following steps. The dimensions and joints of a human arm were determined and reproduced as a robotic arm, where software was designed to run simulations of the robotic arm to make corrections before the final prototype design was produced. The robotic arm was implemented according to the specifications obtained and a motor control circuit was produced to replicate each of the seven movements of the robotic arm. Finally, a validation was performed for each of the movements performed by the robotic arm by considering the position, speed of flexion, and extension of the joints.

1 Introduction

The signals produced by the arm muscles of people who experience amputation can operate a prosthesis or a robotic device. At present, no advanced prototypes have been demonstrated in amputees but there have been many contributions, including:

U.G. Zapana · R.M.C. Apaza · N.I.O. Ordoñez (✉) · A.C. Rivera
Applied Bioengineering Institute, National University of St Agustin of Arequipa, Arequipa, Perú
e-mail: norihue@yahoo.com

R.M.C. Apaza
e-mail: renee.condori@abi-unsa.edu.pe

U.G. Zapana
e-mail: ulises_gordillo@abi-unsa.edu.pe

A.C. Rivera
e-mail: alfredo2429@hotmail.com

A. Kecskeméthy and F. Geu Flores (eds.), *Interdisciplinary Applications of Kinematics*, Mechanisms and Machine Science 26, DOI 10.1007/978-3-319-10723-3_16

KNU Hand from the Intelligent Robot Laboratory in Korea (Englehart et al. 2001), which operates in four zones using muscle signal to execute nine movements with one hand; Shadow Dextrous Hand from the Shadow Robot Company (López et al. 2009), which uses a CyberGlove to acquire finger and wrist positions that are transmitted to a piston-controlled robot; and the method produced by Cabinet Argentina Medical Technology, which is an example of an actual application to people with motor dysfunctions (Betancourt et al. 2004). In the present study, we also aimed develop to new tools for myoelectric prostheses based on current technology.

The project was initiated with the future aim of industrializing the product, thus we started by designing and implementing Ag/AgCl electrodes to acquire myoelectric signals, which we combined with a differential amplifier instrumentation to obtain an electromyography (EMG) system. The captured signals were processed by a digital signal processor using the Fourier wavelet transform and statistical parameters to recognize patterns based on neural networks, thereby improving the degree of certainty for the movements made by robotic arm. Finally, we designed and implemented a robotic arm with anthropomorphic characteristics, which reproduced seven of the main movements of a human arm.

2 Methodology and Implementation

This project was divided into three main themes, where each contributed to the overall goal.

First, we developed modules to capture muscle signals, where we designed and implemented a system of EMG electrodes that comprised actual surface instrumentation amplifier circuits and voltage level adaptation. To capture the electrical pulses, we designed Ag/AgCl surface contact type electrodes, which we also produced using an electrodeposition process. These electrodes were chemical transducers that produced an electrical signal, which was proportional to the chemical parameter. We also manufactured a potassium chloride gel that helped to improve the acquisition of myoelectric signals. Figure 1 shows the manufactured electrode, which is available on the market.

Fig. 1 Ag/AgCl electrodes

2.1 Instrumentation Amplifier

The EMG signals were <10 mV, thus to develop prototype upper limb prosthetics controlled by myoelectric signals, it was necessary to perform differential amplification, thereby eliminating environmental noise or noise due to other muscles (e.g., the heart). The circuit comprised a differential amplifier AD620 shown in Fig. 2. We can also use INA118 instrumentation amplifier with the same purpose.

We designed a circuit that could be adapted to different noisy environments and muscle areas, where the selection of capacitors allowed greater sensitivity in patients with low muscle strength and endurance by calibrating the desired amplification. In the present study, we used a 100 μF capacitor and a resistance of 10 ohms, which provided us with 200-fold amplification (Fig. 3).

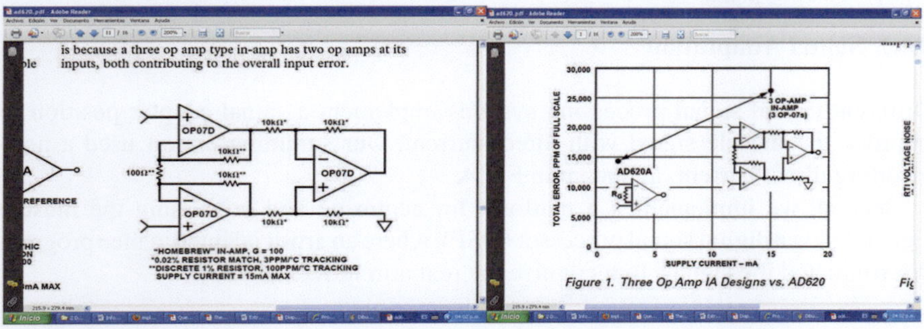

Fig. 2 Differential amplifier

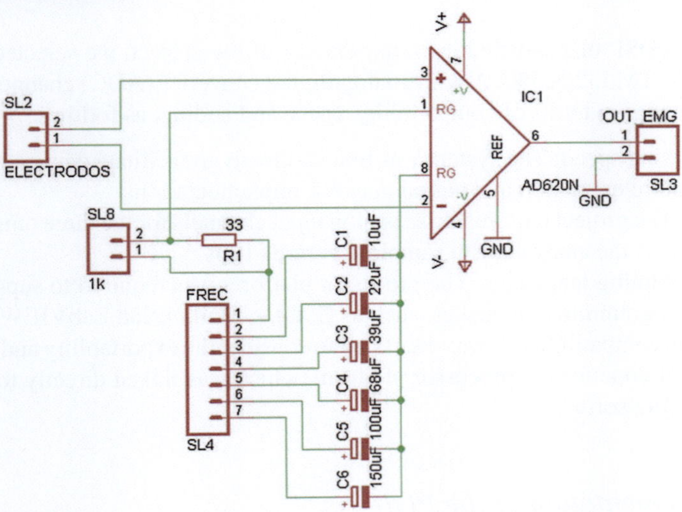

Fig. 3 Circuit that can be adapted to different noisy environments

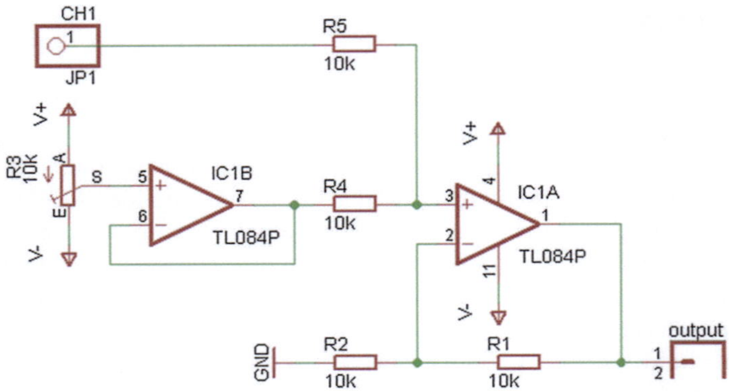

Fig. 4 Signal adaptation circuit

2.1.1 Signal Adaptation

Different digital signal processing systems implement a signal adapter position to improve the muscle signal with direct current. Our summing circuit used a non-inverting direct current, as shown in Fig. 4.

Second, we implemented a platform for acquiring and processing the muscle signals using a digital signal processor (DSP), where an artificial intelligence program discriminated the signals to recognize the real arm movements.

2.2 DSP Specifications

To obtain a DSP that satisfied the requirements of the project, we selected a Texas Instruments TMS320C5509A with analog/digital converter (ADC) channels, which had advantages in terms of compatibility, costs, and coding, as follows.

(a) Processing speed: The system was linked directly to facilitate data management for feature extraction and neural network implementation.
(b) ADC: The project required three analog input channels for the three muscle areas to support the analysis with sampling periods 1 ms.
(c) Programming languages: The processor platform was required to support standard programming languages, such as C, C++, Matlab, and LabVIEW.
(d) Coding compatibility: It was necessary to consider the exportability and importability of code for the processor platform, which was linked directly to the programming software.

2.3 Implementation of the Platform

Care was taken to select the appropriate tool from the market, and the development platform is shown in Fig. 5.

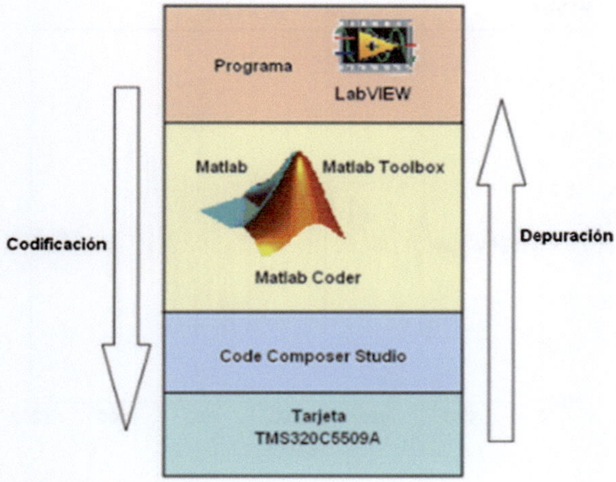

Fig. 5 Software implementation

2.4 Feature Extraction

According to Englehart, the first 400 ms of a muscle movement (Romo et al. 2007) are sufficient for identifying the movement, thus the extraction procedure used a signal amplitude window that started when the movement was detected by thresholding. Feature extraction is an important part of the pattern recognition process. For EMG signals, a pattern is represented by a signal x(t) in the time domain, as shown in Fig. 6.

In general, the time signals have a limited duration, and they are sampled and converted into digital format. It is more appropriate to describe a pattern as a finite time sequence: x [1], x [2], …, x[n]. However, submitting this sequence directly to the classifier is impractical because of the large number of entries and the randomness of the signal. Therefore, the sequence x (n) should be reduced to a vector x = (x1, x2, …, xd), where $d \ll n$, which is called the feature vector, characterizes the signal and allows stable classifier training, thus classification is achieved with a good success rate. In the present study, the features were extracted using a temporal approach, which comprised autoregressive parametric modeling, fast Fourier transform, sliding window short-time Fourier transform, and wavelet transform. This process generated a hybrid vector of features that allowed the identification of EMG signals.

2.4.1 Signal Adaptation

From (Farfan et al. 2005), it is known that a signal acquired by an EMG surface electrode can be considered as the output of a lowpass filter that represents the muscle and the muscle fibers, where the filter output is approximately Gaussian.

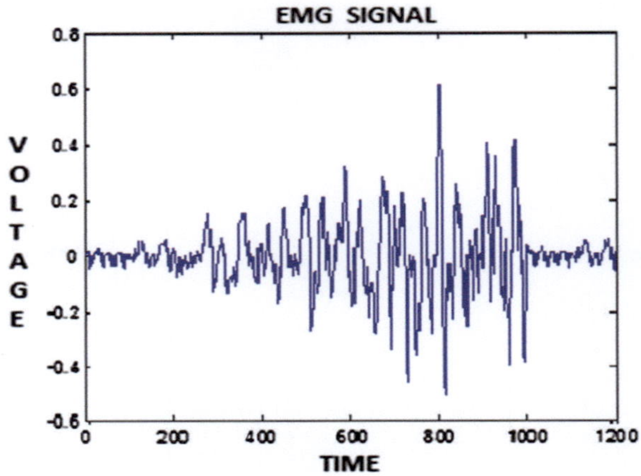

Fig. 6 EMG signals

The optimal filter for a Gaussian process is a linear filter, which may be selected from the AR model, moving average (MA), and ARMA. However, due to the low computational costs of AR (Guyton et al. 2006), this model was used to represent the best approximation of the EMG signal in the present study.

AR Model

Coefficients of the AR model. The coefficients employed in the parametric AR model contain information about the location of the signal peaks in its spectrum. The raw EMG signal obtained from the surface electrodes was modeled as a stochastic time series. The AR is a linear stationary model. The EMG signal is neither linear nor stationary, thus the AR model was considered for short time intervals (100 to 500 ms), according to (Guyton et al. 2006). Thus, EMG analysis could be performed based on the AR model.

The AR model is defined as:

$$x\,(n) = \sum_{i=1}^{P} a_{i+1} x\,(n-i) + e\,(n) \tag{1}$$

where, x (n) is the EMG signal, a_i are the coefficients, P denotes the order of the model, and e (n) is an error term.

Approach with Wavelet Packets

The Fourier transform of a signal f (t) is generally defined by:

$$F(\omega) = \int_{-\infty}^{\infty} f(t) e^{-j\omega t} dt \tag{2}$$

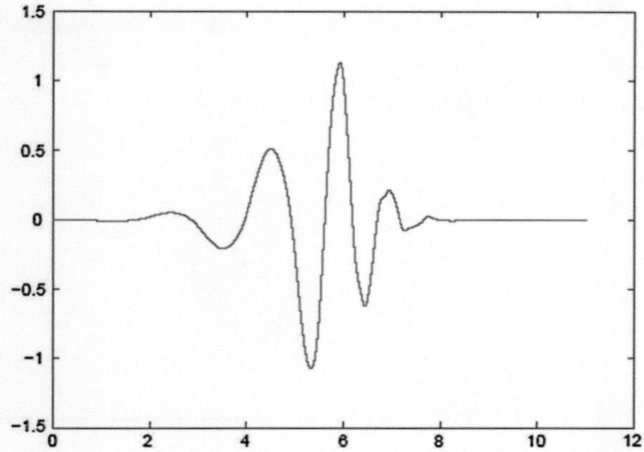

Fig. 7 Daubechies 6th order mother wavelet

where f (t) is measured over a finite time interval T and sampled each time interval τ. The Fourier transform can also be viewed as a bank of N filters (where $\mathrm{N} = \mathrm{T}/\tau$). The filters cover the frequency region $[-\pi/\tau,\ \pi/\tau]$ each bandwidth of $\Delta\omega = 2\pi/\mathrm{T}$. The energy distribution f (t) over the frequency range is:

$$|F(\omega)|,\ \mathrm{y} = \int_{-\pi/\tau}^{\pi/\tau} |F(\omega)|^2 d\omega$$

Note that F (ω) does not contain any information over time.

A wavelet transform was developed to maintain both properties of the temporal frequency of the signal. The analysis was based on a complete set of functions (called wavelets), which could be located in both the time domain and the frequency. The wavelet transform f (t) was defined in terms of two parameters, a and b, where a represents the frequency scale and b indicates the location in time of the wavelet (Khushaba and Al-Juamily 2007).

The Daubechies 6th order mother wavelet used, as shown in Fig. 7. For a signal f (t) measured over a finite time interval T and sampled each time interval τ, the wavelet transform can also be viewed as a bank of discrete filters (Englehart et al. 2001).

2.5 Robotic Arm Measures

The design and implementation of a robot arm required the establishment of baseline measures and materials. To specify the lengths, we measured the phalanges (proximal, intermediate, and distal) of the fingers and the hand using a specially designed

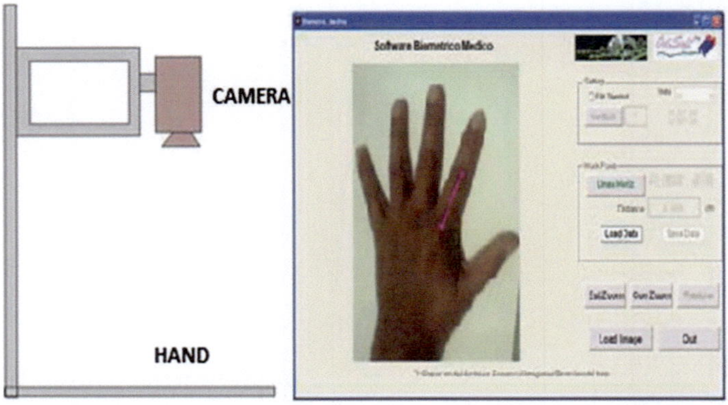

Fig. 8 Camera and medical biometric software

camera support and we employed medical biometric software in MATLAB (Fig. 8), where we obtained measurements from 10 people (Table 1).

2.6 Prototype Design

The robot arm prototype was designed using Inventor software. Further details for one finger are shown in Fig. 9.

The thumb, index, and middle fingers included touch sensors to allow contact sensitivity with objects. This sensor included a switch that activated when touching an object. The complete hand comprised five fingers with two servomotors that operated at 1.5 kg/cm, where the first controlled four fingers and the second controlled the thumb.

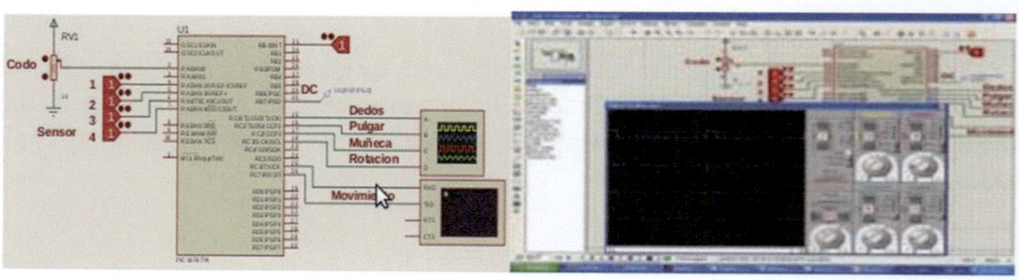

Fig. 9 System motor control

Table 1 Different hand measures

Nro	Thumb			Index			Middle			Ring			Little finger			Palm	
	Phalanges (cm)			Phalanges (cm)			Phalanges (cm)			Phalanges (cm)			Phalanges (cm)			(cm)	
	Proximal	Intermediate	Distal	Proximal	Intermediate	Distal	Proximal	Intermediate	Distal	Proximal	Intermediate	Distal	Proximal	Intermediate	Distal	Width	Length
1	3.25	3.16	3.05	4.1	2.84	2.32	4.23	3.91	2.33	4.15	3.9	2.34	3.95	2.8	2.25	8.51	9.51
2	3.27	3.18	3.07	4.12	2.86	2.34	4.25	3.93	2.35	4.17	3.92	2.36	3.97	2.82	2.27	8.53	9.53
0	3.26	3.17	3.06	4.11	2.85	2.33	4.24	3.92	2.34	4.16	3.91	2.35	3.96	2.81	2.26	8.52	9.52
4	3.28	3.19	3.08	4.13	2.87	2.35	4.26	3.94	2.36	4.18	3.93	2.37	3.98	2.83	2.28	8.54	9.54
5	3.24	3.15	3.04	4.09	2.83	2.31	4.22	3.9	2.32	4.14	3.89	2.33	3.94	2.79	2.24	8.5	9.5
6	3.25	3.16	3.05	4.1	2.84	2.32	4.23	3.91	2.33	4.15	3.9	2.34	3.95	2.8	2.25	8.51	9.51
7	3.23	3.14	3.03	4.08	2.82	2.3	4.21	3.89	2.31	4.13	3.88	2.32	3.93	2.78	2.23	8.49	9.49
8	3.26	3.17	3.06	4.11	2.85	2.33	4.24	3.92	2.34	4.16	3.91	2.35	3.96	2.81	2.26	8.52	9.52
9	3.28	3.19	3.08	4.13	2.87	2.35	4.26	3.94	2.36	4.18	3.93	2.37	3.98	2.83	2.28	8.54	9.54
10	3.22	3.13	3.02	4.07	2.81	2.29	4.2	3.88	2.3	4.12	3.87	2.31	3.92	2.77	2.22	8.48	9.48
Promedio	3.254	3.164	3.054	4.104	2.844	2.324	4.234	3.914	2.334	4.154	3.904	2.344	3.954	2.804	2.254	8.514	9.514

2.6.1 Implementation of Motor Control

We designed a motor control system and microcontroller-based sensor, which comprised four servo systems with three touch sensors on the thumb, index, and middle fingers to limit the arm movement when holding objects in the hand.

In addition, a DC motor with a potentiometer feedback-based flexion and extension forearm received the motor control command via serial communication using the USART Pattern Recognition System in the DSP.

2.6.2 Myoelectric Signal Capture

The muscle signals were captured using the Ag/AgCl electrodes, which obtained the signal shown in Fig. 10.

2.6.3 Feature Extraction

The features were extracted (Fig. 11) from seven movements: open and closing the hand, hand adduction and abduction, pronation and supination of the hand, and forearm flexion and extension, where we analyzed 10 replicates of each movement.

2.6.4 Training Neural Networks

One of the most important steps in this process was the training of a supervised learning neural network, i.e., a multilayer perceptron, using the backpropagation

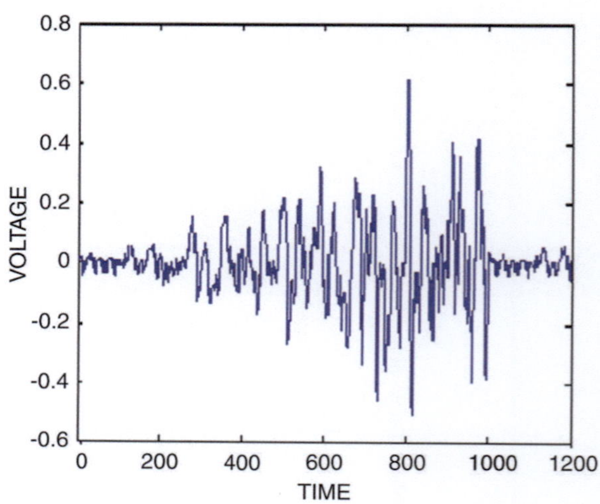

Fig. 10 Myoelectric signal

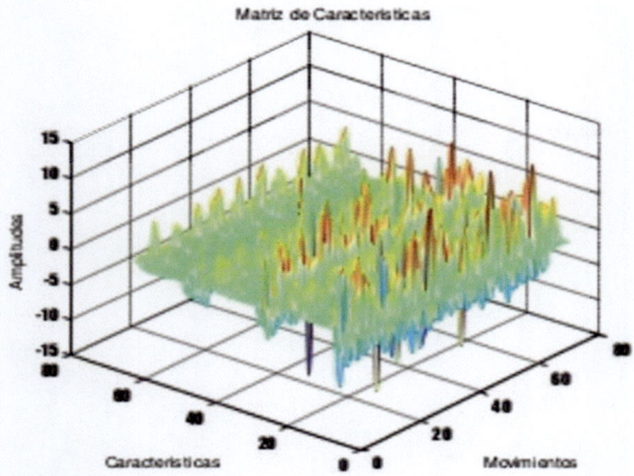

Fig. 11 Features extracted from seven movements

algorithm. The backpropagation neural network has various configuration options, which depend on the learning requirements and the application being developed.

The command in the MATLAB Neural Network Toolbox for implementing RNA is newff to create a backpropagation network, which determines the training algorithm parameters.

In this algorithm, the learning network stopped if the number of iterations exceeded the net.trainParam.epochs command, if the error value reached set a target, if the gradient magnitude was less than net.trainParam.min_grad, or if the training time exceeded the net.trainParam.time. We used a four-layer network with 60, 600, 49, and seven neurons in each layer, where the activation function used for all layers was 'tansig' with values from –1 to 1, and the training algorithm was 'traingdx.' After training, the data were stored in 'net.mat,' 'stdp.mat,' and 'meanp.mat' for use in later stages.

Finally, in the third stage, we developed an anthropomorphic robot arm that reproduced the movements recognized during training.

3 Implementation of the Robot Arm

The robot arm was developed according to the Inventor software design (Fig. 12) based on the average biometric measurements for the arm, hand, and fingers.

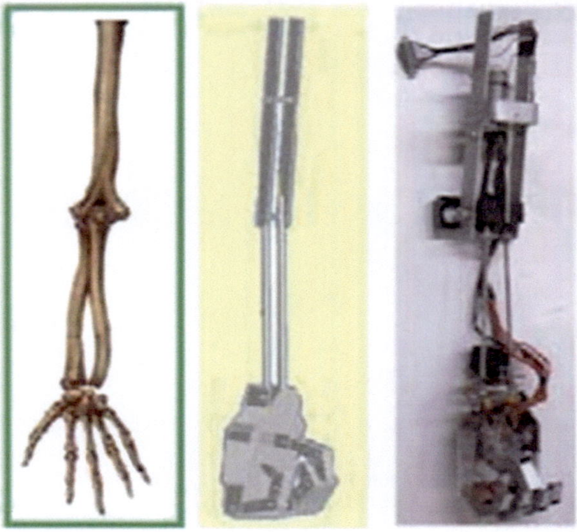

Fig. 12 Morphological comparison of a human arm, the designed arm, and the implemented robot arm

4 Conclusions

In this study, EMG signals were acquired using modern techniques to allow memory storage, where the analysis was performed by a DSP using the entire signal and there was no data loss during this procedure, thereby ensuring its reliability.

We designed and implemented an anthropomorphic robot arm, which replicated the basic movements of the human arm. We considered the skeletal characteristics to generate axes of motion and we incorporated accurate power servomotors that simulated the fibers and muscle groups.

A model was implemented using images extracted from accurate measurements of the hand, finger, and arm joints, which provided us with the basic dimensions for the robotic arm. The information obtained by the camera was sufficient to obtain valid readings.

Hybrid techniques were used by the classifier to obtain better performance, which was demonstrated by the training time for the neural network and the percentage success rate. Compared with previous studies that used the same number of electrodes, the percentages of correct classification using hybrid features vectors were 5 to 10 % higher with our method.

References

Acosta M (2000) Tutorial sobre redes neuronales aplicadas en Ingeniera eléctrica y su implementación. Universidad Tecnológica de Pereira

Betancourt G, Suárez E, Franco, J (2004) Reconocimiento de patrones de movimiento a partir de señales electromiografícas. In: Scientia et Technica Año X, vol 53, no 26. UTP. ISSN 0122–1701

Du S, Vuskovic M (2003) Temporal vs. spectral approach to feature extraction from prehensile EMG signals. Department of Computer Science, San Diego State University, San Diego

Englehart K (1998) Signal representation for classification of the transient myoelectric signal. Doctoral Thesis

Englehart K, Hudgins B, Parker P (2001) A wavelet-based continuous classification scheme for multifunction myoelectric control. IEEE Trans Biomed Eng 48(3):302–311

Farfan F, Polittiy J, Felice C (2005) Evaluación de patrones temporales y espectrales para el control Mioeléctrico. XV Congreso de Bioingeniería, Argentina

Guyton A, John M, Hall E (2006) Textbook of Medical Physiology. Elsevier, Philadelphia

Khushaba R, Al-Juamily A (2007) Fuzzy wavelet packet based feature extraction method for multifunction myoelectric control. Int J Biomed Sci 2(3)

Kuo S, Lee B (2010) Real-time digital signal processor. Implementations, applications and experiments with TMS320C55x. In: Nilsj N (ed) The quest for artificial intelligence: a history of ideas and achievements. Stanford University

López N, di Sciascio F, Soria C, Valentinuzzi M (2009) Robust EMG sensing system based on data fusion for myoelectric control of a robotic arm. Facultad Ingeniería, Universidad Nacional de San Juan, Gabinete de Tecnología Médica, Argentina

Romo H, Realpe EJ, Jojoa P (2007) Análisis de Señales EMG Superficiales y su Aplicación en Control de Prótesis de Mano PhD. Universidad del Cauca

Shadow Robot Company (2011) http://www.shadowrobot.com/

A 3D Foot-Ground Model Using Disk Contacts

Matthew Millard and Andrés Kecskeméthy

Abstract A foot contact model is an important component of any forward-dynamic human gait simulation. This work presents a preliminary experimental validation of a three-dimensional (3D) foot contact model that represents the heel and forefoot using a pair of contact disks. The disk elements are well-suited to modeling the foot because they are computationally efficient and are mechanically stable when flat on the ground. We evaluated the foot model by comparing its ankle position to the subject's ankle position (measured using skin-mounted reflective markers and infrared cameras) when both feet developed the same ground reaction force (GRF) and center-of-pressure (COP) profiles (measured using a force plate). We used this novel approach because the experimental GRF and COP measurements are accurate, but the kinematic data is usually corrupted with 1 cm of skin-movement error at the foot. The results indicate that the disk-based foot model is an accurate representation of the subject's barefoot except during toe-off, and when the COP is on the extreme medial boundaries of the foot. The experimental data and foot model presented in this work is provided as supplementary material online.

1 Introduction

Human motion prediction *in-silico* is of great interest to many research communities because of its potential to improve our understanding of healthy and pathological locomotion. During normal bipedal locomotion feet are the only contacts that interact with the ground. A validated model of the human foot is thus a pre-requisite for a computed prediction of human gait.

A predictive gait simulation requires a foot model that can simultaneously reproduce the kinematics, ground reaction forces (GRF) and center-of-pressure (COP) profile of a human foot. In addition, a 3D foot contact model must be as stable

M. Millard (✉) · A. Kecskeméthy
Universität of Duisburg-Essen, Duisburg, Germany
e-mail: matthew.millard@uni-due.de

A. Kecskeméthy
e-mail: andres.kecskemethy@uni-due.de

A. Kecskeméthy and F. Geu Flores (eds.), *Interdisciplinary Applications of Kinematics*, Mechanisms and Machine Science 26, DOI 10.1007/978-3-319-10723-3_17

as a human foot, so that the ankle does not roll unrealistically during simulation. Although the foot can be modeled using realistic geometry (Seth et al. 2011), this is computationally expensive. Instead, most foot contact models use fast, but simplified contact-pairs to represent the foot: point-plane (Scott and Winter 1998; Peasgood et al. 2007; Ackermann and Bogert 2010; Gilchrist and Winter 1996; Neptune et al. 2000; Wilson et al. 2006), circle-plane (2D) (Millard et al. 2009), sphere-plane (Pàmies-Vilà et al. 2012; Cuadrado et al. 2011), and disk-plane (3D) (Kecskeméthy 2011). Here we focus on the disk-based 3D foot contact model (Kecskeméthy 2011) because the disk contact approximates the round shape of the foot pads, is mechanically stable, and is computationally efficient.

Though the foot has been modeled using a variety of approaches, relatively few foot-ground contact models have been tested against experimental data (Gilchrist and Winter 1996; Millard et al. 2009; Kecskeméthy 2011). Conceptually a foot contact model transforms a kinematic state into a wrench applied at the ankle, and vice-versa (though the mapping from wrench to kinematic state is not unique). A good validation method matches one side of this transformation to experimental data, and then evaluates the model against experimental data using the other side of the transformation.

The development of well-posed, well-behaved foot contact validation methods is an area of active research. Skin-movement artifact precludes using one-to-one experimental kinematics as input to the foot contact model and then evaluating it by comparing simulated GRF and COP profiles to experimental data (Millard et al. 2009): the 1 cm of skin movement error at the foot (Holden et al. 1997) is comparable to the expected heel (Gefen et al. 2001) and metatarsal pad compressions (Cavanagh 1999). Numerical instability precludes applying the observed wrench to the model's ankle, forward integrating, and then evaluating the model by comparing simulated foot kinematics to experimental kinematics (Gilchrist and Winter 1996): applying large forces to a body of low mass results in a simulation that is very sensitive to initial conditions and experimental error. This approach can be vastly improved by ignoring the dynamics of the foot, solving for a position of static equilibrium, and comparing the resulting foot orientation to experimental data (Kecskeméthy 2011) though the resulting root-finding problem is not easily solved. In this paper we present a novel control system which drives the ankle of the foot model to a trajectory that reproduces the accurately measured GRF and COP profiles. We evaluate the model by comparing its ankle kinematics to experimental observations: distance errors of 1 cm (Holden et al. 1997) or less are acceptable, since this amount of skin-movement error is present in motion capture data.

The experimental tasks that are used to validate the foot contact model are just as important as the validation methods. Although walking data is commonly used to evaluate foot contact models (Gilchrist and Winter 1996; Millard et al. 2009; Kecskeméthy 2011), additional motions should be used because the kinematics of the foot and its COP profile during walking are mostly in the sagittal plane. We have recorded a novel trial where the subject moves the COP of their barefoot in the pattern of a growing spiral which eventually traces the outer edge of their foot.

The spiraling-COP task allows the foot model to be tested against motions in both the sagittal and frontal planes, providing a more comprehensive evaluation of the foot-ground contact model.

2 Foot Model

The 3D foot contact model is composed of a hind foot, and a forefoot body joined together by a revolute joint. The axis of this joint has been rotated 22° about the $\hat{z}'$ axis to match the raked alignment of the subject's metatarsal joints. The musculature of the toes is represented with a nonlinear spring-damper element in parallel with the revolute joint. The heel contact is represented by a disk (Kecskeméthy et al. 2000) attached to the hind foot, while the metatarsal and toe contacts are represented by a single disk attached to the forefoot body (Fig. 1).

The disk interacts with the ground plane through a point contact (Kecskeméthy et al. 2000). The point contact lies in the direction

$$ {}_M\hat{\mathbf{u}}_P = \hat{\mathbf{n}}_C \times (\hat{\mathbf{n}}_C \times \hat{\mathbf{n}}_P) \tag{1}$$

which points from the center of the disk to point P, the lowest point on the disk edge. This direction vector is computed using the surface normal vector $\hat{\mathbf{n}}_C$ and the disk normal vector $\hat{\mathbf{n}}_P$. The contact point is located along ${}_M\hat{\mathbf{u}}_P$ at a distance of

$$ r = r_0(1 - e^{-C \sin\alpha}) \tag{2}$$

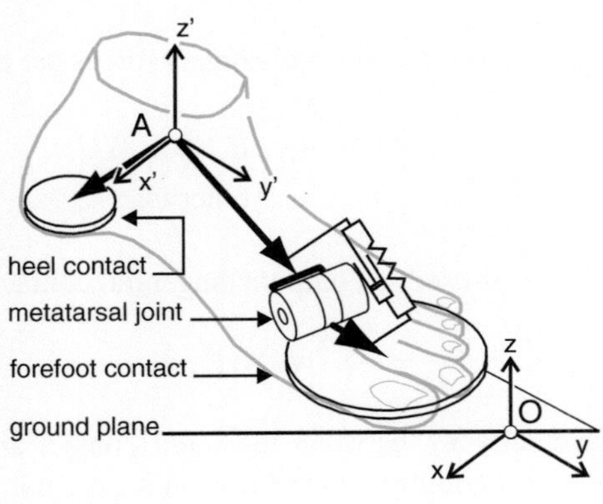

Fig. 1 The foot interacts with the ground plane through two disk contacts. The heel disk is rigidly attached the ankle frame (*frame A*). A revolute metatarsal joint allows the fore-foot disk to rotate with respect to the heel disk. A spring-damper across the metatarsal joint applies a torque to the ankle frame. The *left foot* is shown because the subject's left foot was instrumented

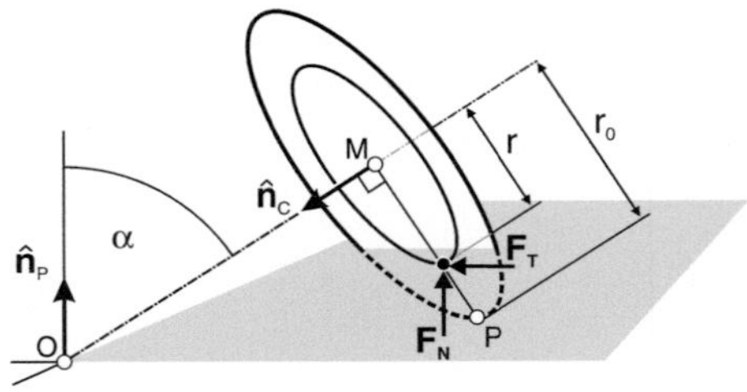

Fig. 2 The contact point of the disk model (Kecskeméthy et al. 2000), shown in *black*, is smoothly brought to the center of the disk as it flattens (as $\alpha \to 0$). The smooth transition makes the contact kinematics of the disk similar to those of an infinitely thin rounded surface. The rounded appearance of the disk lends itself to simulating human foot pads, which have a *flat bottom* and *rounded edges*

from the center of the disk, where α is the angle between the surface and disk normal vectors. An exponential function makes the disk appear rounded by smoothly blending r from 0, when the disk is flat on the ground ($\alpha = 0$), to approach the disk radius r_0, when the disk is perpendicular to the ground plane ($\alpha = \frac{\pi}{2}$). The parameter C is used to control how rounded the disk appears. Low values of C will make the disk edges appear more rounded, while high values of C will make the disk appear sharper (Fig. 2).

Once the point contact location has been computed, the state of the disk contact is used to calculate the penetration depth, x_N, and velocity, v_N, of the point contact into the surface. Here we use a Hunt-Crossley contact model (Hunt and Crossley 1975) to compute the normal force magnitude

$$F_N^* = k x_N^P (1 + d v_N) \tag{3}$$

with the condition that only positive normal contact forces are permitted, i.e.

$$\mathbf{F}_N = \begin{cases} F_N^* \cdot \hat{\mathbf{n}}_P & \text{if } F_N^* > 0 \\ 0 & \text{otherwise.} \end{cases} \tag{4}$$

A Coulomb friction model is used to compute tangential contact forces

$$\mathbf{F}_T = -\mu(|\mathbf{v}_T|_2) F_N \, \hat{\mathbf{v}}_T \tag{5}$$

using the tangential velocity $\mathbf{v}_T$ between the point contact and the surface. The coefficient of friction μ is smoothly interpolated (using a cubic spline) to make the system equations less stiff. We compute the tangential velocity direction, $\hat{\mathbf{v}}_T$, using the numerically stable method described in Eq. 20 of Gonthier et al. (2004).

3 Methods

We evaluated the foot contact model using experimental foot kinematic and kinetic data collected during a conventional barefoot walking task, and a novel task designed to move the COP over the entire bottom surface of the foot in the pattern of a growing spiral. The COP-spiral began with COP rotations about the subject's mid-foot, and grew over 5 rotations to the outside of the subject's foot while under a load of $\approx$3/4 bodyweight (the subject held onto a support to maintain balance during this task). Foot kinematics were measured using infrared cameras and skin-mounted reflective markers, while ground reaction forces and moments were measured using a floor-mounted force plate.

To evaluate the foot contact model we computed the Euclidean distance between the subject's ankle position $\tilde{\mathbf{r}}_A(t)$ and the model's ankle position $\mathbf{r}_A(t)$ (Fig. 1)

$$\varepsilon(t) = ||\tilde{\mathbf{r}}_A(t) - \mathbf{r}_A(t)|| \tag{6}$$

when both feet generated the same GRF and COP profiles (experimental quantities indicated using the tilde symbol). In the context of this error measure, an excellent foot contact model will be able to remain within a skin-movement distance ($\varepsilon \leq 1$ cm (Holden et al. 1997)) of the subject's ankle while accurately reproducing the observed GRF and COP profiles. We used this novel evaluation method because it allows the foot model to be evaluated accurately in the presence of skin-movement error.

A control system and forward-simulation was used to compute the path of the foot model that reproduced the experimental GRF and COP profiles. The control system consists of three parts: a feed-forward wrench computed from the experimental data $\tilde{\mathbf{W}}$, a kinematic feedback proportional-derivative (PD) tracking controller ($[\mathbf{P}]\mathbf{q}_e + [\mathbf{D}]\dot{\mathbf{q}}_e$), a force feedback controller $[\mathbf{E}]\mathbf{W}_e$, and a damping wrench $[\mathbf{V}]\dot{\mathbf{q}}$:

$$\mathbf{W}_u = \tilde{\mathbf{W}} + (1 - s)\, C_Q\, ([\mathbf{P}]\mathbf{q}_e + [\mathbf{D}]\dot{\mathbf{q}}_e) + s\, C_W\, [\mathbf{E}]\mathbf{W}_e + s\, C_V [\mathbf{V}]\dot{\mathbf{q}}. \tag{7}$$

Here a subscript e, as in q_e of Eq. 7, is used to indicate an error, defined as the difference between an experimental observation and the corresponding value in the simulation. All of the gain matrices in Eq. 7 are diagonal, and have been tuned by hand. The tracking, force-feedback, and damping control terms in Eq. 7 are scaled by task dependent scalar gain variables C_Q, C_W, and C_V respectively. These gains are tuned so that the model's GRF and COP profiles quickly and smoothly converge to the experimentally measured profiles during the walking and spiral tasks. A blending variable

$$s = \frac{\tilde{F}_N(t)}{\max(\tilde{F}_N(t))} \tag{8}$$

is used to ensure the kinematic tracking controller dominates when contact forces are low, and the force feedback controller dominates when contact forces are high.

The results of a forward simulation are used to compute $\varepsilon(t)$ in Eq. 6, the distance between the subject's ankle and the model's ankle.

The parameters of the foot model were tuned using the distance error ε to guide changes. The geometric parameters of the foot model were initially matched as closely as possible to the subject's foot using manual measurements. The stiffness of the contact disks were set so that the heel and metatarsal pads compress by approximately 50 % of their unloaded thickness during stance (Gefen et al. 2001). For this subject, a 50 % compression of the heel pad amounts to 11.4 (22.8 mm thick), and 6.7 mm (13.2 mm thick) for the 2nd metatarsal pad. The thickness of the subject's foot pads were measured previously using an ultrasound scanner. The stiffness and damping of the toe joint was initially manually set. A combination of derivative-free optimization and hand tuning was used to adjust the geometry of the foot and the stiffness of the toe joint to find a single set of parameters that fit the experimental data taken from the walking and spiraling-COP tasks.

4 Results and Discussion

The walking trial shows that the model foot is indistinguishable from the subject's foot until toe-off. The distance error, ε, between the model's ankle and the subject's ankle is less than the experimental skin-stretch error of 1 cm (Holden et al. 1997) until $t = 0.39$ s (Panel a of Fig. 3). An association between error and the foot model is made in Panel b of Fig. 3 by plotting the COP projected on the model foot, using a line that becomes thicker and darker as the distance error, ε, increases. The thickest

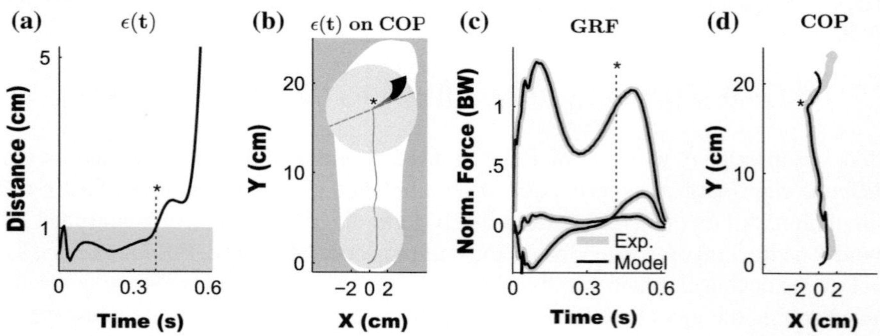

Fig. 3 Panel **a** illustrates the distance error, $\varepsilon(t)$, between the model's ankle and the subject's ankle during the walking experiment. The error exceeds 1 cm at $t = 0.39$ s, which is marked with an '*.' Panel **b** illustrates the COP (as seen from above) on the surface of the model's foot, using a line that becomes *thicker* and *darker* as the distance error, $\varepsilon(t)$, increases. The *dashed line* shows the axis of rotation of the metatarsal joint (n.b. *left* foot is shown). To provide context, the outside of the subject's foot is shown in *white*, and the disk contacts are shown in *light gray*. Panels **c** and **d** illustrate the GRF and COP profiles of the model (*black line*) and the experiment (*grey line*)

and darkest lines of the COP error plot (Panel b of Fig. 3) show that the distance error is small until the COP extends past the metatarsal joint (Panel b of Fig. 3), the equivalent location of the subject's big toe. The control system drives the model to accurately track the experimental GRF and COP profiles with little error, except during heel contact and toe-off (Panels c and d of Fig. 3). The COP error that occurs during initial heel contact ($t = 0$ s) is caused by a shape mismatch between the heel disk and the subject's heel in the initial pose. Due to this shape mismatch we could not simultaneously collocate both the ankle and the COP at the start of the stance phase. We chose to match the position of the ankle, introducing an error in the COP profile, so that the two ankles would continue to align as the foot rotated and flattened on the ground plane. The COP error at toe-off occurs because the contact disk does not extend far enough to cover the contact patch of the big toe.

The spiraling-COP trial results show that the model foot closely tracks the experimental data for the first 3 COP rotations (4.6 s). The error steadily accumulates during the entire trial, we suspect, due to small differences in the shape of the model's foot and the subject's foot. The distance error, ε, can be associated with a location on the foot model in Panel b of Fig. 3 by plotting the COP projected onto the surface of the model foot using a line that becomes thicker and darker as the distance error ε increases. At the beginning of the trial, the COP-error line is light and thin, but steadily grows with each subsequent rotation (Panel b of Fig. 4). During the final rotation, both the distance error ε and the COP tracking error become large as the COP traverses the subject's toe, and proceeds down the medial side of the subject's left foot. The COP was on the extreme edge of the subject's foot, which was everted by 21° (relative to the ground plane), during this final movement.

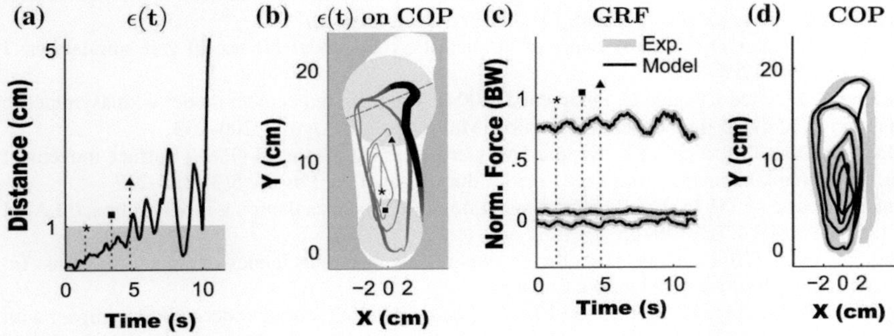

Fig. 4 Panel **a** illustrates the distance error, $\varepsilon(t)$, between the model ankle and the human ankle during the spiraling-COP task. The first 3 rotations are marked using '*', ■, and ▲ symbols. Panel **b** illustrates the COP (as seen from *above*) on the surface of the model's foot, using a line that becomes *thicker* and *darker* as the distance error, $\varepsilon(t)$, increases. The *dashed line* shows the axis of rotation of the metatarsal joint of the *left* foot. To provide context, the outside of the subject's foot is shown in *white*, and the disk contacts are shown in *light gray*. Panels **c** and **d** illustrate the GRF and COP profiles from the model (*black line*) and the experiment (*grey line*)

The high level of agreement between the model and experimental data during these two different tasks is very encouraging. In both tasks, the model matches the data within experimental error when the COP is on the heel, or is between the heel and metatarsal joint. It is only when the COP is beyond the metatarsal joint, or when the foot is everted by 21° (about the ground plane), that the model's ankle kinematics differ substantially from the experimental data. The error between the model and the experimental data during the final loop of the spiral task is acceptable for most simulations since it required an extreme tilt of 21°. The close agreement during the first 3 rotations of the spiraling-COP task suggests that the model's heel and mid-foot is as resistant to ankle rolling as the subject's foot. Further efforts will concentrate on improving the fidelity of the toe, and characterizing the model's stability.

Acknowledgments The authors gratefully acknowledge Dr. Dominik Raab for his generous assistance with the data collection. This work was co-funded by the German federal state North Rhine Westphalia (NRW) and the European Union (European Regional Development Fund: Investing In Your Future).

References

Ackermann M, van den Bogert AJ (2010) Optimality principles for model-based prediction of human gait. J Biomech 43(6):1055–1060

Cavanagh P (1999) Plantar soft tissue thickness during ground contact in walking. J Biomech 32:623–628

Cuadrado J, Pàmies-Vilà R, Lugrís U, JF Alonso (2011) A force-based approach for joint-efforts estimation during the double support phase of gait. In: Procedia IUTAM, vol 2. Waterloo, Canada

Gefen A, Megido-Ravid M, Itzchak Y (2001) In vivo biomechanical behavior of the human heel pad during the stance phase of gait. J Biomech 34:1661–1665

Gilchrist L, Winter D (1996) A two-part viscoelastic foot model for use in gait simulations. J Biomech 29(6):795–798

Gonthier Y, McPhee J, Lange C, Piedbœuf J (2004) A regularized contact model with asymmetric damping and dwell-time dependent friction. Multibody Syst Dyn 11:209–233

Holden JP, Orsini JA, Siegel KL, Kepple TM, Gerber LH, Stanhope SJ (1997) Surface movement errors in shank kinematics and knee kinetics during gait. Gait Posture 5(3):217–227

Hunt K, Crossley F (1975) Coefficient of restitution interpreted as damping in vibroimpact. J Appl Mech-T ASME 42(E):440445

Kecskeméthy A (2011) Integrating efficient kinematics in biomechanics of human motions. In: Procedia IUTAM, vol 2. Waterloo, Canada

Kecskeméthy A, Lange C, Grabner G (2000) A geometric model for cylinder-cylinder impact with application to vertebrae motion simulation. In: Lenarčic J, Stanišić M (eds) Advances in Robot Kinematics. Springer, Netherlands, pp 345–354

Millard M, McPhee J, Kubica E (2009) Multi-step forward dynamic gait simulation. In: Bottasso C (ed) Multibody Dynamics: Computational Methods and Applications. Springer, New York, pp 25–43

Neptune RR, Wright IC, van den Bogert AJ (2000) A method for numerical simulation of single limb ground contact events: application to heel-toe running. Comput Methods Biomec 3:321–334

Pàmies-Vilà R, Font-Llagunes J, Lugrís U, Cuadrado J (2012) Two approaches to estimate foot-ground contact model parameters using opitmization techniques. In: IMSD. Stuttgart

Peasgood M, Kubica E, McPhee J (2007) Stabilization of a dynamic walking gait simulation. J Comput Nonlinear Dyn 2:65–72
Scott S, Winter DA (1998) Biomechanical model of the human foot: kinematics and kinetics during the stance phase of walking. J Biomech 26:1091–1104
Seth A, Sherman M, Reinbolt JA, Delp SL (2011) OpenSim: a musculoskeletal modeling and simulation framework for in silico investigations and exchange. In: Procedia IUTAM, vol 2. Waterloo, Canada
Wilson C, King MA, Yeadon MR (2006) Determination of subject-specific model parameters for visco-elastic elements. J Biomech 39(10):1883–1890

Fitting Useful Planar Four-Bar and Six-Bar Linkages to Over-Specified Tasks

Brandon Y. Tsuge and J. Michael McCarthy

Abstract In this paper, we introduce a method to design useful four-bar and six-bar linkages that reach more than the five specified task positions used in Burmester synthesis of RR cranks. Similar to existing methods, we minimize the crank-length error in each task position, however our approach does not introduce penalty functions to eliminate defective linkage. Instead, we search user defined tolerance zones around the task positions to find defect-free, or usable, linkage designs. Finally, candidate linkages are ranked by average link length. A numerical example is provided for seven task positions for a four and six-bar linkage.

1 Introduction

This paper presents a synthesis procedure for usable four-bar and six-bar linkages that reach an over-specified set of task positions, often termed approximate motion synthesis, Ravani and Roth (1983). This approach combines the optimization-based method of Shen et al. (2009) with the randomized adjustment of the task position introduced by Plecnik and McCarthy (2011), and results in a number of usable linkages that are ranked by their link lengths. The use of tolerance zones expands the number of linkage candidates.

The design procedure for planar four-bar linkages that guide a coupler link through five task positions is known as Burmester theory (McCarthy and Soh 2010). The constraint equations of a four-bar linkage can be solved exactly for five task positions, while approximation approaches are needed for more than five positions.

Recent examples of various optimization approaches to linkage synthesis, including the use of penalty functions, that ensure defect-free linkages and minimum

B.Y. Tsuge (✉) · J.M. McCarthy
University of California, Irvine, USA
e-mail: btsuge@uci.edu

J.M. McCarthy
e-mail: jmmccart@uci.edu

A. Kecskeméthy and F. Geu Flores (eds.), *Interdisciplinary Applications of Kinematics*, Mechanisms and Machine Science 26, DOI 10.1007/978-3-319-10723-3_18

transmission angles can be found in research by Sancibrian (2011), Alizade et al. (2013) and Penunuri et al. (2012).

In the study of exact synthesis of linkages (Stumph and Murray 2000; Holte et al. 2000; Mlinar and Erdman 2000) focused on the relatively imprecise nature of the designer's specification of the task. It is often the case that the designer is satisfied with all of the positions within a tolerance zone around a specific task position. This leads to the use of tolerance zones in the exact synthesis algorithm presented by Plecnik and McCarthy (2012), which finds a defect free 5-SS linkage.

This paper explores the use of tolerance zones to find usable linkage that reach an over-specified set of task positions.

2 Synthesis for Over-Specified Task Positions

Burmester theory provides synthesis equations that can be solved exactly to define RR cranks used to construct four-bar and six-bar linkages, Fig. 1a, b. The designer specifies a set of coordinate transformations, $[T_j], j = 1, \ldots, N$, that define positions of the coupler link in the ground frame, called task positions,

$$[T_j] = \begin{bmatrix} \cos\theta_j & -\sin\theta_j & x_j \\ \sin\theta_j & \cos\theta_j & y_j \\ 0 & 0 & 1 \end{bmatrix} \qquad i = 1, \ldots, N, \tag{1}$$

where $\mathbf{d}_j = (x,_j\, y_j)$ define the origin of the moving frame M and θ_j define its orientation.

It is convenient to formulate the task positions as relative displacements, D_{1j} $j = 2, \ldots, N$, from the first task position, given by

$$[D_{1j}] = [T_j][T_1]^{-1}, \quad j = 2, \ldots, N. \tag{2}$$

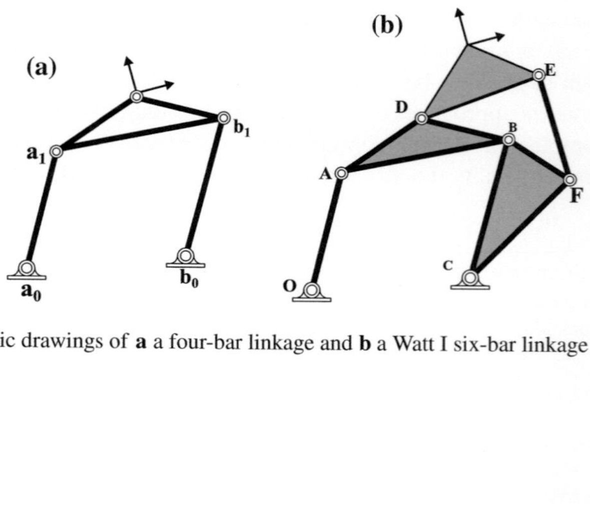

Fig. 1 Schematic drawings of **a** a four-bar linkage and **b** a Watt I six-bar linkage

The requirement that the moving pivots $\mathbf{a}_1 = (a_{1x}, a_{1x}, 1)^T$ and $\mathbf{b}_1 = (b_{1x}, b_{1x}, 1)^T$ lie on a circle relative to the fixed pivots $\mathbf{a}_0 = (a_{0x}, a_{0x}, 1)^T$ and $\mathbf{b}_0 = (b_{0x}, b_{0x}, 1)^T$ in all of the task positions yields the constraint equations,

$$
\begin{aligned}
&([D_{1j}](\mathbf{a}_1 - \mathbf{a}_0) \cdot ([D_{1j}](\mathbf{a}_1 - \mathbf{a}_0) - (\mathbf{a}_1 - \mathbf{a}_0) \cdot (\mathbf{a}_1 - \mathbf{a}_0) = 0,\\
&([D_{1j}](\mathbf{b}_1 - \mathbf{b}_0) \cdot ([D_{1j}](\mathbf{b}_1 - \mathbf{b}_0) - (\mathbf{b}_1 - \mathbf{b}_0) \cdot (\mathbf{b}_1 - \mathbf{b}_0) = 0.
\end{aligned} \tag{3}
$$

These equations can be solved to determine as many as six, four-bar linkages that reach the five task positions.

The equations provided by Burmester theory cannot be solved exactly if the number of task positions is greater than five. However, an approximate solution can be obtained by minimizing the crank-error over all of the over-specified number of task positions. Sum Eqs. (2) and (3) to define the crank error function,

$$
\begin{aligned}
E = \sum_{j=2}^{N} \Big\{ &\big[([D_{1j}]\mathbf{a}_1 - \mathbf{a}_0)^T ([D_{1j}]\mathbf{a}_1 - \mathbf{a}_0) - (\mathbf{a}_1 - \mathbf{a}_0)^T (\mathbf{a}_1 - \mathbf{a}_0)\big]^2 \\
&+ \big[([D_{1j}]\mathbf{b}_1 - \mathbf{b}_0)^T ([D_{1j}]\mathbf{b}_1 - \mathbf{b}_0) - (\mathbf{b}_1 - \mathbf{b}_0)^T (\mathbf{b}_1 - \mathbf{b}_0)\big]^2 \Big\}.
\end{aligned} \tag{4}
$$

This function is the sum of the squares of the differences in crank lengths in each task position. It measures the variation of the crank length needed to reach more than five task positions.

We used the Mathematica Levenberg-Marquardt function to yield values of the fixed and moving pivots in the first task position that minimize this error function. For 7 positions, this minimization algorithm obtained results in less than 1 min.

3 Sorting Solutions

The solutions resulting from the minimization problem are filtered for duplicates and defects. The first defect occurs when minimization of E converges to the same crank. This is a due to the starting positions causing convergence to the same local minimum. The next step to the defect removal process involves the removal of duplicates or solutions that are nearly duplicates of each other. Solutions are considered to be duplicates if the following condition exists.

$$\|[a_{0k}, a_{1k}, b_{0k}, b_{1k}] - [a_{0k}, a_{1k}, b_{0k}, b_{1k}]\| < e \tag{5}$$

where e is a user defined constant that determines how close two solutions need to be before they are considered the same. In addition, linkages that are considered to have a defect if its linkage perimeter is deemed to be too large, as shown in Eq. (6)

$$L = l_1 + l_2 + l_3 + l_4 \tag{6}$$

where l_1, l_2, l_3, and l_4 are the link lengths. L is specified by the user. Once defective solutions are removed, the remaining solutions are ranked based on the Euclidean norm of the link lengths. This was done so that linkages with a larger average size were ranked lower, than ones with a relatively small average size. The linkage candidates were also ranked by the ratio of the smallest link length to the largest link length, but it was found that the ranking of the linkage solutions didn't change dramatically.

4 Adjusted Task Positions

The coordinates and the orientation angle of the task positions can be adjusted within a specified tolerance zone. These adjusted task positions are then run through the same minimization and sorting algorithm to yield additional solutions. The adjustment step can be repeated a number of times in order to increase the number of solutions. The solutions from the original solutions are then inserted into the list of solutions from the adjusted task positions; each of the solutions from the original task positions are inserted into the new list based on the norm of the link lengths. Figure 2b shows the small variation between the original task positions and the adjusted task positions.

The four-bar linkage synthesis method used in this paper can be used to design Watt I six-bar Linkages, Fig. 1b. Soh and McCarthy (2008) introduced a method to design this six bar linkage by defining a 3R chain and synthesizing two four-bar linkages. We used our four-bar synthesis method combined with the six-bar linkage synthesis method to design a linkage with seven task positions.

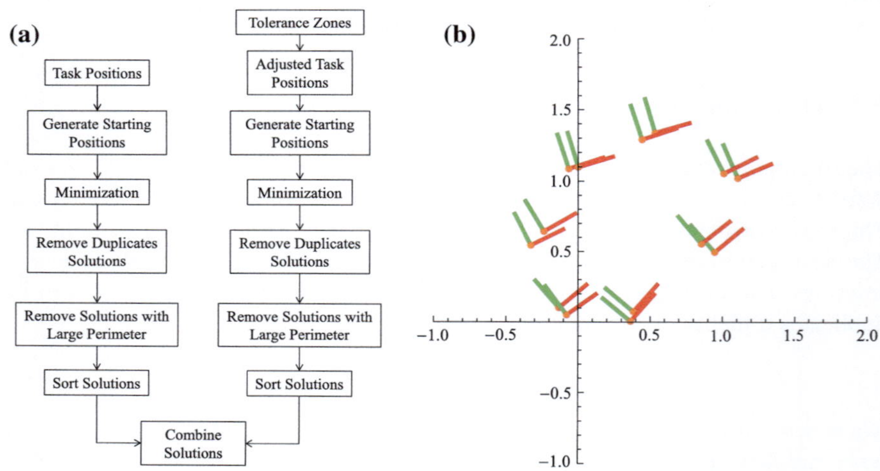

Fig. 2 The design flow chart and an example of the original tasks and an adjusted set of tasks selected from within the tolerance zones. **a** Design flow chart, **b** Original and adjusted tasks

Table 1 Seven task positions

Task	x	y	θ
1	0.53	1.33	16.04
2	−0.07	1.09	18.13
3	−0.33	0.54	26.87
4	−0.13	0.10	40.40
5	0.38	0.07	49.16
6	0.94	0.50	40.55
7	1.10	1.02	23.14

5 Four-Bar Linkage Synthesis

We demonstrate our design methodology by finding usable four-bar linkages that guide the coupler through seven task positions. The tasks positions are shown in Table 1, and the sorted linkage design candidates are shown in Table 2.

Figure 3a, b shows the resulting linkages and their crank error. After ranking the linkages by the norm of the link lengths, solutions four and five were not considered to be useful, due to their large average area.

The tasks positions are adjusted by selecting new task positions within the specified tolerance zones, and the design process is repeated. In this case, it is executed ten times. The x and y coordinates of the task positions were adjusted within tolerance of ± 0.1 and $\pm 0.1°$. This yielded 50 linkage designs, which were sorted by average link length.

Figure 4a shows solutions 18 and 19 that have crank error $E = 0.89$ and 0.25, which is equivalent to the crank error of solution 1 for the original task positions. Figure 4b shows solutions 20 and 21, that have crank-error values $E = 0.45$ and 0.67, which are equivalent to the solution 2 for the original task positions.

Lastly, the linkage solutions that resulted from the adjusted task positions had crank error that ranged from 0.01 to 118.27. There were also 33 solutions that had crank error that was less than the smallest crank error value from the solutions resulting from the original task positions.

6 Synthesis of a Watt-I Six-Bar Linkage

The same task positions used in the previous, four-bar linkage example was used to synthesize a usable six-bar linkage. The 3R serial chain was specified to have a fixed pivot at $(-1.25, -1.25)$, and all of the link lengths were set to 1.25, as shown in Fig. 5a.

The four-bar synthesis algorithm for the two four-bar linkages contained in the six-bar linkage resulted in two different solutions and their crank error values are

Table 2 Design candidates

Solution	a_{0x}	a_{0y}	a_{1x}	a_{1y}	b_{0x}	b_{0y}	b_{1x}	b_{1y}
1	-2.46×10^{-3}	-4.6×10^{-5}	-2.69×10^{-4}	0.75	2.59	1.50	2.70	4.6×10^{-3}
2	2.70	$4.6(10^{-4})$	2.59	1.50	5.42	0.89	5.16	1.61
3	-2.46×10^{-4}	-4.6×10^{-5}	-2.69×10^{-4}	0.75	5.42	0.89	5.16	1.61
4	-5.35×10^{-4}	-9.2×10^{-5}	-5.65×10^{-4}	0.45	2.55	9.16	5.11	-8.10
5	0.66	2.35	1.22	2.84	2.55	9.15	5.11	-8.10

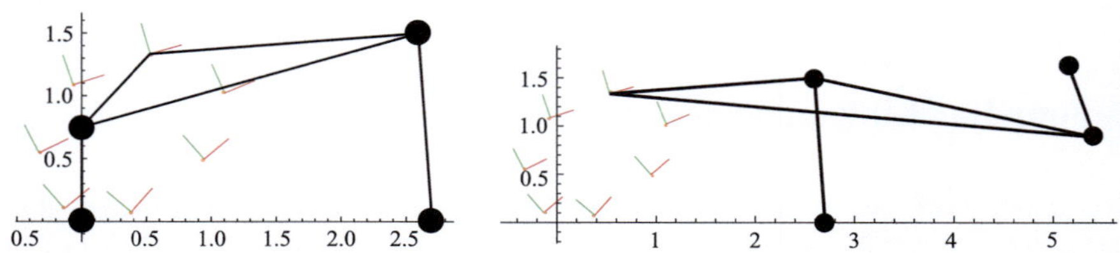

(a) Solution 1. **(b)** Solution 2.

Fig. 3 Linkage solutions with crank error values of **a** $E = 2.75 \times 10^8$, and **b** $E = 0.22$

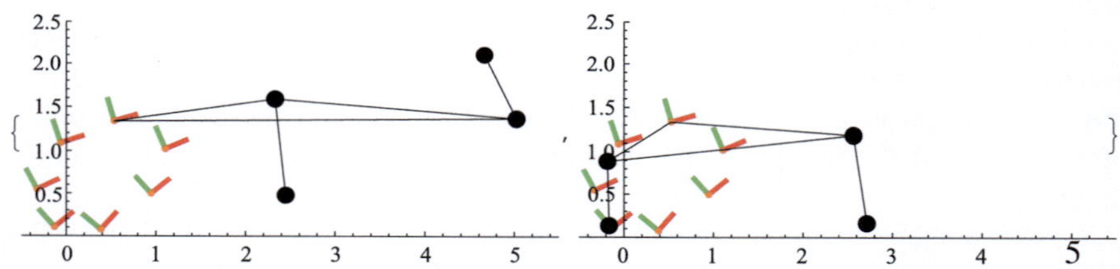

(a) Solution 18 and 19.

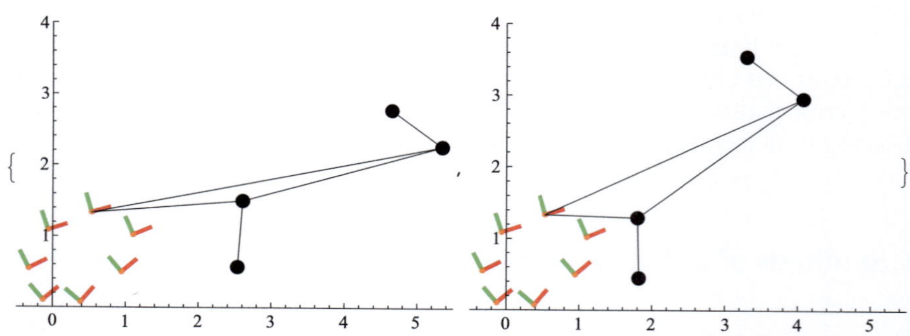

(b) Solution 20 and 21.

Fig. 4 Linkage solutions from adjusted task positions with crank error values of **a** $E = 0.89$ and 0.25, and **b** $E = 0.45$ and 0.67

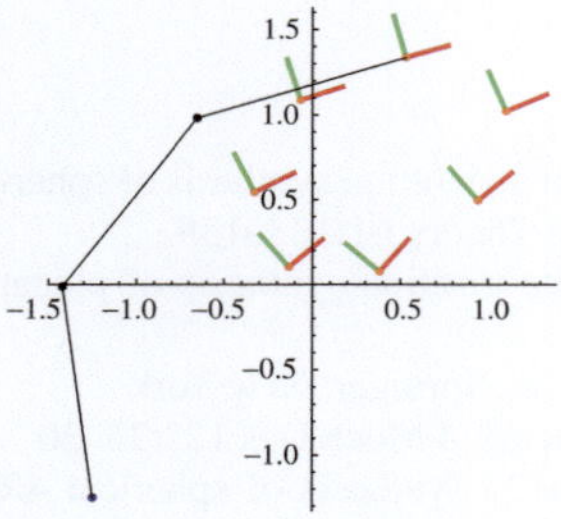

(a) 3R serial chain

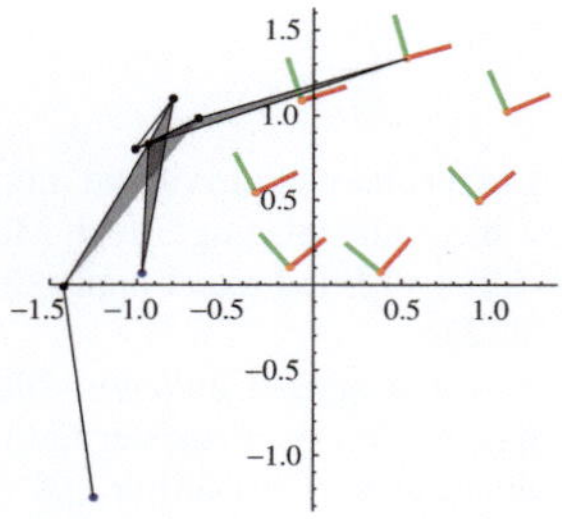

(b) Six-bar solution 1

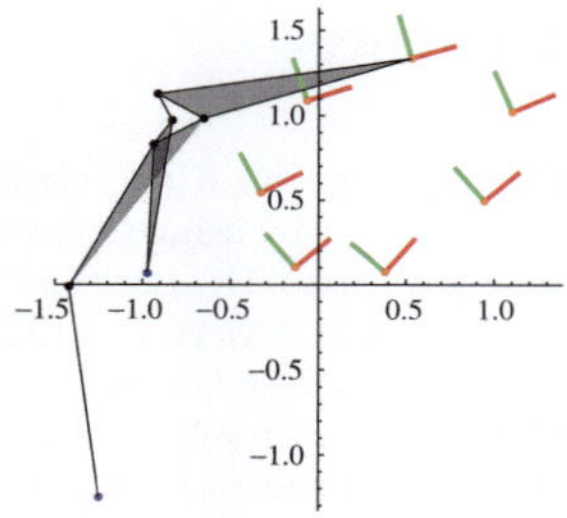

(c) Six-bar solution 2

Fig. 5 The 3R serial chain and the two six-bar linkage solutions crank error of **a** $E = 4.99 \times 10^{-3}$ and $E = 2.40 \times 10^{-3}$, and **b** $E = 4.99 \times 10^{-3}$ and 7.4×10^{-4}

Table 3 Points **O**, **A**, **B**, **C**, **D**, **E**, **F** defining six-bar solutions

Solution	O_x	O_y	A_x	A_y	B_x	B_y	C_X	C_y	D_x	D_y	E_x	E_y	F_x	F_y
1	−1.25	−1.25	−1.42	−0.01	−0.94	0.83	-0.97	0.07	−0.65	0.98	−1.01	0.80	−0.80	1.10
2	−1.25	−1.25	−1.42	−0.01	−0.94	0.83	−0.97	0.07	−0.65	0.98	−0.91	1.13	−0.83	0.97

shown in Fig. 5. The coordinates of the pivots of the two six-bar linkages that resulted from the method are shown in Table 3.

The task positions were adjusted with the same tolerance zones as the previous four-bar linkage, numerical example. The algorithm was looped 10 times and a total of 19 solutions were attained. The error for the first four-bar linkage ranged from 5.80×10^{-3} to 0.36 and the error of the second four-bar linkage varied from 1.92×10^{-3} to 1.71. Since these E values are relatively small and close to the values from the non-adjusted task positions, varying the task coordinates and orientation angle slightly can provide additional linkage designs.

7 Conclusion

In this paper, we present a design methodology for four-bar and six-bar linkages that reach an over-specified set of task positions, which means there is no exact Burmester solution. Our method uses tolerance zones around the task positions to adjust the $N > 5$ task specification. We demonstrate the method by designing a four-bar and a six-bar linkage to reach a seven task positions.

By adjusting the task positions within user specified tolerance zones, our design procedure yielded 33 additional four-bar linkage candidates that achieved a crank error similar to that obtained for the original set of task positions. In the case of six-bar linkage synthesis, adjusting the task positions yielded an additional 19 candidates with a similar crank error. This approach expands the number of candidate linkages that approximate a given over-specified task.

References

Alizade R, Can FC, Kilit O (2013) Least square approximate motion generation synthesis of spherical linkages by using Chebyshev and equal spacing. Mech Mach Theory 61:123–135

Holte JR, Chase TR, Erdman AG (2000) Mixed exact-approximate position synthesis of planar mechanisms. J Mech Des 122:278–286

McCarthy JM, Soh GS (2010) Geometric design of linkages, 2nd edn. Springer, New York

Mlinar R, Erdman AG (2000) An introduction to burmester field theory. J Mech Des 122:25–30

Penunuri F, Peon-Escalante R, Villanueva C, Cruz-Villar CA (2012) Synthesis of spherical 4R mechanism for path generation using differential evolution. Mech Mach Theory 57:62–70

Plecnik MM, McCarthy JM (2011) Five position synthesis of a Slider-Crank function generator. In: Proceedings of the ASME 2011 international design engineering technical conferences and computers and information in engineering conference, Paper No. DETC2011-4, Washington, 28–31 Aug 2011

Plecnik MM, McCarthy JM (2012) Design of a 5-SS spatial steering linkage. In: Proceedings of the ASME 2012 international design engineering technical conferences, Paper No. DETC 2012–71405, Chicago, 12–15 Aug 2012

Ravani B, Roth B (1983) Motion Synthesis Using Kinematic Mappings. ASME J Mech Transm Autom Des 105:460–467

Sancibrian R (2011) Improved CRG method for the optimal synthesis of linkages in function generation problems. Mech Mach Theory 46:1350–1375

Shen Q, Al-Smadi YM, Martin PJ, Russell K, Sodhi RS (2009) An extension of mechanism design optimization for motion generation. Mech Mach Theory 44:1759–1767

Soh GS, McCarthy JM (2008) The synthesis of six-bar linkages as constrained planar 3R chains. Mech Mach Theory 43(2):160–170

Stumph HE, Murray AP (2000) 11 Defect-free Slider-Crank function generation for 4.5 precision points. In: Proceedings of the ASME 2000 design engineering techni-cal conference, Paper DETC2000/MECH-14070, Baltimore, 10–13 Sept 2000

On the Requirements of Interpolating Polynomials for Path Motion Constraints

Jorge Ambrósio, Pedro Antunes and João Pombo

Abstract In the framework of multibody dynamics, the path motion constraint enforces that a body follows a predefined curve being its rotations with respect to the curve moving frame also prescribed. The kinematic constraint formulation requires the evaluation of the fourth derivative of the curve with respect to its arc length. Regardless of the fact that higher order polynomials lead to unwanted curve oscillations, at least a fifth order polynomials is required to formulate this constraint. From the point of view of geometric control lower order polynomials are preferred. This work shows that for multibody dynamic formulations with dependent coordinates the use of cubic polynomials is possible, being the dynamic response similar to that obtained with higher order polynomials. The stabilization of the equations of motion, always required to control the constraint violations during long analysis periods due to the inherent numerical errors of the integration process, is enough to correct the error introduced by using a lower order polynomial interpolation and thus forfeiting the analytical requirement for higher order polynomials.

Keywords Multibody dynamics · Constraint violations · Moving frames · Constraint stabilization

J. Ambrósio (✉) · P. Antunes
LAETA, IDMEC, Instituto Superior Técnico, University of Lisbon, Lisbon, Portugal
e-mail: jorge@dem.ist.utl.pt

P. Antunes
e-mail: pedro.antunes@live.com

J. Pombo
LAETA, IDMEC, Instituto Superior Técnico, University of Lisbon, Lisbon, Portugal
e-mail: jpombo@dem.ist.utl.pt

J. Pombo
ISEL, Instituto Politécnico de Lisboa, Lisbon, Portugal

A. Kecskeméthy and F. Geu Flores (eds.), *Interdisciplinary Applications of Kinematics*, Mechanisms and Machine Science 26, DOI 10.1007/978-3-319-10723-3_19

1 Introduction

The definition of railway, tramway or roller coaster tracks requires the accurate description of their geometries, which is usually done with the parameterization of the track centerline. The definition of a reference plane where the rails sit is required and consequently a curve moving frame must also be specified.

Regardless of the definition of a general curve geometry being to specify a railway centerline, a rollercoaster spatial geometry or a prescribed motion kinematic constraint not only a suitable interpolation scheme must be selected but also a robust definition of the curve moving frame needs to be used. Depending on if the curve is used to set some geometric layout for the mechanical models or to define kinematic constraints for multibody dynamics applications, a minimum order derivative, with respect to the curve parameter, may be required. Therefore, using polynomial interpolation schemes, higher order polynomials may be required for an exact formulation of problem. Generally, higher order interpolating polynomials lead to unwanted, and hardly controllable, oscillations of the curve geometry, i.e., small deviations of the positions of the nodal points lead to large variations of the curve geometry away from those nodes. But although lower order polynomials generally have a local geometric control they may not have the order necessary to ensure the proper geometric continuity of the model or the parametric derivatives required in the formulation of a kinematic constraint. Therefore, the question that arises, for which some light is shed here, is: what are the minimal requirements that an interpolating polynomial must meet in to be used in the definition of a path motion kinematic constraint?

The geometric description of the curve must allow the definition of a moving frame in which the tangent, normal and binormal vectors define an orthogonal frame. Both Frenet and Dabroux frames are candidates to play the role of the required moving frame (Frenet 1852; Darboux 1889). As discussed by Tandl and Kecskemethy, both have singularities in general spatial curve geometries, as those required for rollercoaster analysis (Tandl and Kecskemethy 2006; Tandl 2008). In this work the Frenet frame is used being the straight segments handled with the provision described by Pombo and Ambrosio (2003).

Using the selected moving frame definition a proper formulation for a prescribed motion kinematic constrained is obtained. Such kinematic constraint imposes that a point of a rigid body follows a given curve and that the body itself does not rotate, or does it in a prescribed manner, with respect to the curve moving frame. Depending on the choice of coordinates used on the multibody formulation this kinematic constraint may be defined differently (Nikravesh 1988; Jalon and Bayo 1993). When Cartesian coordinates are used and the equations of motion are solved together with the second time derivative of the position kinematic constraints the definition of the Frenet frame requires the second derivative of the curve with respect to its arc length parameter while the acceleration constraints, i.e., the second time derivative of the kinematic constraint, requires the existence of a fourth derivative. In this sense, apparently fifth order polynomials are required to formulate properly the prescribed motion kinematic constraint.

The numerical integration of the equations of motion of a multibody systems implies the use of numerical integrators, such as Runga-Kutta, Gear or others, to undertake the forward dynamic analysis (Shampine and Gordon 1975; Gear 1971). All numerical procedures used in the solution of the equations of motion and on their solution have a finite precision and ultimately lead to small errors that affect the precision of the solution. When dependent coordinates, such as Cartesian or Natural coordinates, are used only the acceleration constraints are explicitly used in the solution of the equations of motion being the position and velocity constraints fulfilled only if the numerical integration would be error free, being otherwise violated and leading to instabilities in the dynamic solution of the analysis (Nikravesh 1988). By using stabilization procedures, such as the Baumgarte constraint stabilization method (Baumgarte 1972) or the Augmented Lagrangian Formulation (Bayo and Avello 1994) such constraint violations can be kept under control. By using a coordinate partition scheme the constraint violations can be eliminated (Neto and Ambrósio 2003).

This work shows that the same procedures used to stabilize the constraint violations in the integration of the equations of motion of multibody systems formulated with dependent coordinates also allow for the use of interpolation schemes with polynomials that have an order lower than that required for the exact formulation of the prescribed motion kinematic constraints. It is shown that regardless of using interpolation schemes with higher order polynomials the constraint violations still grow to a point in which either stabilization or coordinate partition procedures are required. Furthermore, it is shown here that when constraint stabilizations methods are used there is no observable difference in the constraint violations between interpolating polynomials of higher and lower order, provided that they satisfy the continuity required for the definition of the curve moving frames and for the geometric requirements of the model.

2 Curve Parameterization

The interpolation of complex curve geometries is generally realized by piecewise polynomial schemes that consist of polynomial pieces of the same degree with a prescribed overall smoothness. The input data includes the coordinates of interpolation points and parameters values to control rotations about the tangent vector (Farin 1990). The advantage of these interpolating procedures is that they exhibit local geometrical control, i.e., the variation of the position of a control point only affects the neighborhood of that point, maintaining unchanged the rest of the curve.

2.1 Curve Parameterization

Let the curve be described using a nth order spline segments that interpolate a set of control points defined as Boor (1978)

$$\mathbf{g}(u) = \begin{Bmatrix} x(u) \\ y(u) \\ z(u) \end{Bmatrix} = \mathbf{a}_0 + \mathbf{a}_1\, u + \mathbf{a}_2\, u^2 + \mathbf{a}_3\, u^3 + \cdots + \mathbf{a}_n\, u^n \tag{1}$$

where $\mathbf{g}(u)$ is a point on the curve, u is the local parametric variable and $\mathbf{a}_i$ are the unknown algebraic coefficients that must be calculated using the boundary conditions, i.e., segment end points and tangent vectors (Boor 1978). Although Eq. (1) is generic for any polynomial interpolation, in this work only cubic, cubic shape-preserving and quintic polynomials are considered.

2.2 Curve Moving Frame

The osculating plane, at a given point P on a curve, is the plane of closest contact to the curve in the neighborhood of P, as depicted in Fig. 1. Although there are different available frames definitions for the purpose (Tandl 2008), the Frenet frame, which provides an appropriate curve referential at every point, is used here. Note that in railway or roller coaster dynamics the track model is developed with reference to the planes defined by the curve moving frame, being the osculating plane the one in which the rails are defined.

The unit vectors that characterize the Frenet frame are the unit tangent vector $\mathbf{t}$, the principal unit normal vector $\mathbf{n}$ and the binormal vector $\mathbf{b}$. These vectors, defined in the intersection of the normal, rectifying and osculating planes shown in Fig. 1, are written as Pombo and Ambrosio (2003)

$$\mathbf{t} = \frac{\mathbf{g}^u}{\|\mathbf{g}^u\|};\, \mathbf{n} = \frac{\mathbf{k}}{\|\mathbf{k}\|};\, \mathbf{b} = \tilde{\mathbf{t}}\mathbf{n} \tag{2}$$

where $\tilde{\mathbf{t}}\mathbf{n}$ means a cross product and the auxiliary vector $\mathbf{k}$ given by

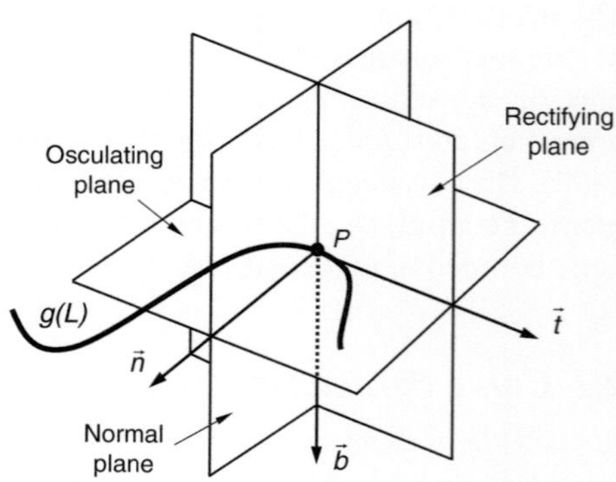

Fig. 1 Principal unit vectors associated to the moving frame of a *curve*

$$\mathbf{k} = \mathbf{g}^{uu} - \frac{\mathbf{g}^{uuT}\mathbf{g}^{u}}{\|\mathbf{g}^{u}\|^{2}}\mathbf{g}^{u} \tag{3}$$

where $\mathbf{g}^u$ and $\mathbf{g}^{uu}$ denote, respectively, the first and second derivatives of the parametric curve $\mathbf{g}(u)$ with respect to the parametric variable u. Note that when piecewise polynomial interpolation is used, a transformation from the curve parameter u to and curve arc-length parameter is required. This issue is discussed with the presentation of the formulation of the prescribed motion constraint.

3 Overview of the Multibody Dynamics Formulation

Let a rigid body i to which a body-fixed reference frame $(\xi,\eta,\zeta)_i$ is rigidly attached be represented in Fig. 2. With Cartesian coordinates, the position and orientation of the rigid body are defined by the translation coordinates $\mathbf{r}_i = \{x, y, z\}_i^T$ and Euler parameters $\mathbf{p}_i = \{e_0, e_1, e_2, e_3\}_i^T$, respectively (Nikravesh 1988). The vector of coordinates associated to the rigid body i is

$$\mathbf{q}_i = \left\{\mathbf{r}^T, \mathbf{p}^T\right\}_i^T \tag{4}$$

A multibody system with nb bodies is described by

$$\mathbf{q} = \left\{\mathbf{q}_1^T, \mathbf{q}_2^T, \ldots, \mathbf{q}_{nb}^T\right\}^T \tag{5}$$

where $\mathbf{q}$ is the vector of generalized coordinates.

In a multibody system, the mechanical joints are used to connect the bodies in order to restrain their relative motions. These kinematic constraints are expressed as algebraic constraint equations that introduce dependencies between the coordinates being denoted by Nikravesh (1988)

$$\boldsymbol{\Phi}(\mathbf{q}, t) = \mathbf{0} \tag{6}$$

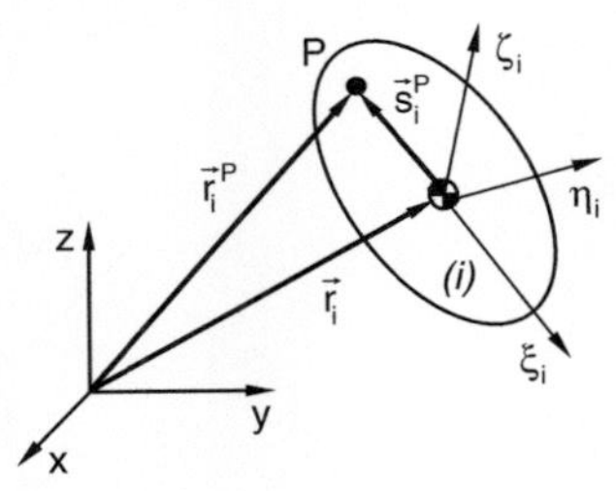

Fig. 2 Rigid body with its body-fixed reference frame

where t is the time. The second time derivative of Eq. (6) yields the acceleration kinematic constraints, or acceleration equations, denoted as

$$\ddot{\Phi}\left(\mathbf{q}, \dot{\mathbf{q}}, \ddot{\mathbf{q}}, t\right) = \mathbf{0} \equiv \Phi_{\mathbf{q}}\ddot{\mathbf{q}} = \boldsymbol{\gamma} \tag{7}$$

where $\Phi_{\mathbf{q}}$ is the Jacobian matrix of the constraints, $\ddot{\mathbf{q}}$ is the acceleration vector and γ is the vector that contains all terms in the equations that are not dependent on the accelerations.

Using this formulation, the equations of motion for a constrained mechanical system can be obtained using the Lagrange multipliers technique as Nikravesh (1988):

$$\mathbf{M}\ddot{\mathbf{q}} + \Phi_{\mathbf{q}}^{T}\lambda = \mathbf{f} \tag{8}$$

where $\mathbf{M}$ is the global mass matrix, containing the mass and moments of inertia of all bodies, λ is the vector of the unknown Lagrange multipliers and $\mathbf{f}$ is the force vector containing all forces and moments applied on system bodies and the gyroscopic forces. Equation (8) represents a system of n second order ordinary differential equations with $n+m$ unknowns, corresponding to the accelerations $\ddot{\mathbf{q}}$ and of the Lagrange multipliers λ. In order to obtain a solution for this equation the m acceleration equations (7) are appended to the equations of motion (8). The resulting equations for a constrained multibody system are rearranged in the matrix form as

$$\begin{bmatrix} \mathbf{M} & \Phi_{\mathbf{q}}^{T} \\ \Phi_{\mathbf{q}} & \mathbf{0} \end{bmatrix} \begin{Bmatrix} \ddot{\mathbf{q}} \\ \lambda \end{Bmatrix} = \begin{Bmatrix} \mathbf{f} \\ \boldsymbol{\gamma} \end{Bmatrix} \tag{9}$$

The dynamic analysis of multibody systems involves the calculation of the vectors $\mathbf{f}$ and γ, for each time step according to the scheme depicted by Fig 3. Equation (9)

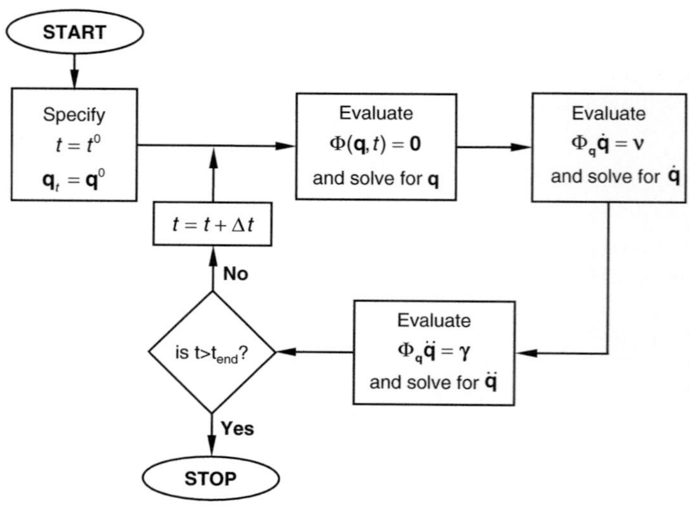

Fig. 3 Flowchart of the computational implementation of a dynamic analysis

is used to calculate the system accelerations $\ddot{\mathbf{q}}$ and the Lagrange multipliers λ. The accelerations together with the velocities $\dot{\mathbf{q}}$ are integrated in order to obtain the new velocities $\dot{\mathbf{q}}$ and positions $\mathbf{q}$ for the next time step. This process proceeds until the final time of the analysis is reached. The Gear integration method (Gear 1971) is used for the numerical integration of the velocities and accelerations.

During long time integrations the numerical errors associated to the numerical integration tend to propagate. As neither the position constraint equations (6) neither their time derivative, the velocity equations, are explicitly used, small constraint violations tend to increase. To keep the process stable, the Baumgarte Stabilization Method (Baumgarte 1972) or the Augmented Lagrangean Formulation (Bayo and Avello 1994) are used, eventually complemented by a coordinate partition method (Neto and Ambrósio 2003).The Baumgarte stabilization method corresponds to the addition of a feedback term to the acceleration equations (7) penalizing position and velocity constraint violations as

$$\Phi_{\mathbf{q}}\,\ddot{\mathbf{q}} = \gamma - 2\alpha\dot{\Phi} - \beta^2\Phi \tag{10}$$

The detailed description of the multibody formulation and numerical methods used here is outside the scope of this work. The interested readers are referred to references (Nikravesh 1988; Jalon and Bayo 1993; Neto and Ambrósio 2003) for further details on the numerical procedures used.

4 Prescribed Motion Kinematic Constraint

The implementation of the prescribed motion constraint in a computer code requires that the piecewise polynomial parameter u is replaced by a curve arc-length parameter L. Consider the parametric variable u^P, corresponding to a point P, located on the kth polynomial segment to which a curve length L_k^P measured from the kth segment origin is associated. The parameter u^P is obtained by Pombo and Ambrosio (2003):

$$\int_0^{u^P} \sqrt{\mathbf{g}_k^{u^T}\,\mathbf{g}_k^u}\,du - L_k^P = 0 \tag{11}$$

where $\mathbf{g}^u$ is the derivative of $\mathbf{g}$ with respect to u. In terms of its computer implementation, the non-linear equation (11) is solved in the program pre-processor using the Newton–Raphson method (Nikravesh 1988). In what follows the references to the derivatives of the interpolating polynomials with respect to the local parameter u are now referred with respect to the length parameter L_k^P written in short as L.

Consider a point P, located on a rigid body i, that is constrained to follow a specified path, as seen in Fig. 4. The path is defined by a parametric curve $g(L)$, which is controlled by a global parameter L that represents the length travelled by

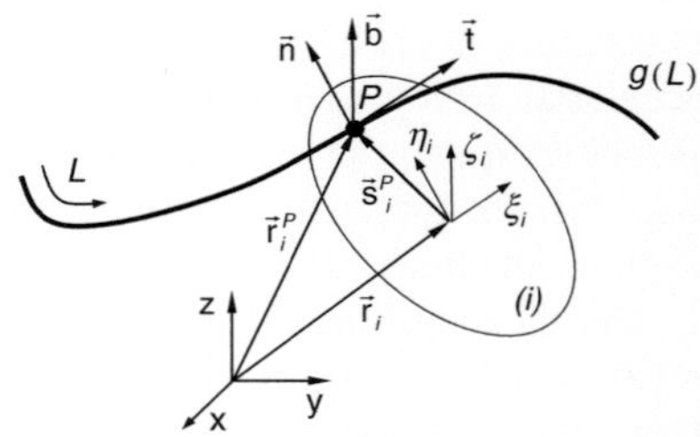

Fig. 4 Prescribed motion constraint

the point along the curve from the origin to the current location of point P. The constraint equations that enforce point P to follow the reference path $g(L)$ are written as Pombo and Ambrosio (2003):

$$\Phi^{(pmc,3)} = \mathbf{0} \equiv \mathbf{r}_i^P - \mathbf{g}(L) = \mathbf{0} \tag{12}$$

where $\mathbf{r}_i^P$ is the position vector associated to point P, depicted in Fig. 4.

The prescribed motion constraint also ensures that the spatial orientation of body i remains unchanged with respect to the moving Frenet frame (**t**, **n**, **b**) associated to the curve. Consider a rigid body i where $(\mathbf{u}_\xi, \mathbf{u}_\eta, \mathbf{u}_\zeta)_i$ represent the unit vectors associated to the axes of $(\xi,\eta,\zeta)_i$ defined in the body frame. Consider also that the Frenet frame of the general parametric curve $\mathbf{g}(L)$ is defined by the principal unit vectors $(\mathbf{t}, \mathbf{n}, \mathbf{b})_L$, as depicted in Fig. 4. The relative orientation between the body vectors $(\mathbf{u}_\xi,\mathbf{u}_\eta,\mathbf{u}_\zeta)_i$ and the curve local frame $(\mathbf{t}, \mathbf{n}, \mathbf{b})_L$ must be such that Pombo and Ambrosio (2003):

$$\Phi^{(lfac,3)} = \begin{Bmatrix} \mathbf{n}^T \mathbf{A}_i \mathbf{u}_\xi \\ \mathbf{b}^T \mathbf{A}_i \mathbf{u}_\xi \\ \mathbf{n}^T \mathbf{A}_i \mathbf{u}_\eta \end{Bmatrix} - \begin{Bmatrix} a \\ b \\ c \end{Bmatrix} = \mathbf{0} \tag{13}$$

where $\{a\ b\ c\}^T = diag\left[\mathbf{A}_L^{0T}\ \mathbf{A}_i^0\right]$ are constants calculated at the initial time of the analysis by using Eq. (13) with the initial conditions.

The contribution of the prescribed motion constraint to the constraint acceleration equations (7) is written as

$$\begin{bmatrix} \Phi_{\mathbf{q}}^{(pmc,3)} \\ \Phi_{\mathbf{q}}^{(lfac,3)} \end{bmatrix} \ddot{\mathbf{q}} = \begin{bmatrix} \boldsymbol{\gamma}^{(pmc,3)\#} \\ \boldsymbol{\gamma}^{(lfac,3)\#} \end{bmatrix} \tag{14}$$

where the Jacobian matrix associated to each part of the kinematic constraint is

$$\Phi_{\mathbf{q}}^{(pmc,3)} = \begin{bmatrix} \mathbf{I} & -\tilde{\mathbf{s}}_i^R \mathbf{A}_i & -d\mathbf{g}/dL \end{bmatrix} \tag{15}$$

$$\boldsymbol{\Phi}_{\mathbf{q}}^{(lfac,3)} = \begin{bmatrix} \mathbf{0}^T & -\mathbf{n}^T\mathbf{A}_i\tilde{\mathbf{u}}_\xi & (d\mathbf{n}/dL)^T\,\mathbf{A}_i\mathbf{u}_\xi \\ \mathbf{0}^T & -\mathbf{b}^T\mathbf{A}_i\tilde{\mathbf{u}}_\xi & (d\mathbf{b}/dL)^T\,\mathbf{A}_i\mathbf{u}_\xi \\ \mathbf{0}^T & -\mathbf{n}^T\mathbf{A}_i\tilde{\mathbf{u}}_\zeta & (d\mathbf{n}/dL)^T\,\mathbf{A}_i\mathbf{u}_\zeta \end{bmatrix} \tag{16}$$

and the contribution of each part of the kinematic constraint to the right hand side of the acceleration equations is

$$\boldsymbol{\gamma}^{(pmc,3)\#} = -\;\tilde{\boldsymbol{\omega}}_i\dot{\mathbf{s}}_i^R + \frac{d^2\mathbf{g}}{dL^2}\,\dot{L}^2 \tag{17}$$

$$\boldsymbol{\gamma}^{(pmc,3)\#} = -\begin{Bmatrix} \left[2\,\dot{L}\,(d\mathbf{n}/dL)^T\,\mathbf{A}_i\tilde{\boldsymbol{\omega}}_i' + \mathbf{n}^T\mathbf{A}_i\tilde{\boldsymbol{\omega}}_i'\tilde{\boldsymbol{\omega}}_i' + \dot{L}^{\,2}\left(d^2\mathbf{n}/dL^2\right)^T\mathbf{A}_i\right]\mathbf{u}_\xi \\ \left[2\,\dot{L}\,(d\mathbf{b}/dL)^T\,\mathbf{A}_i\tilde{\boldsymbol{\omega}}_i' + \mathbf{b}^T\mathbf{A}_i\tilde{\boldsymbol{\omega}}_i'\tilde{\boldsymbol{\omega}}_i' + \dot{L}^{\,2}\left(d^2\mathbf{b}/dL^2\right)^T\mathbf{A}_i\right]\mathbf{u}_\xi \\ \left[2\dot{L}\,(d\mathbf{n}/dL)^T\,\mathbf{A}_i\tilde{\boldsymbol{\omega}}_i' + \mathbf{n}^T\mathbf{A}_i\tilde{\boldsymbol{\omega}}_i'\tilde{\boldsymbol{\omega}}_i' + \dot{L}^{\,2}\left(d^2\mathbf{n}/dL^2\right)^T\mathbf{A}_i\right]\mathbf{u}_\zeta \end{Bmatrix} \tag{18}$$

To understand the minimum requirements for the degree of the interpolating polynomials that can be used in the formulation of the prescribed motion constraint, the order of the derivatives used in Eqs. (12) through (18) must be identified. The right hand side vector in Eq. (18) involves $d^2\mathbf{n}/dL^2$ being $\mathbf{n} = \mathbf{k}/\,\|\mathbf{k}\|$ given by Eq. (2) and $\mathbf{k} = \mathbf{g}^{uu} - \left(\mathbf{g}^{uuT}\mathbf{g}^u\right)\mathbf{g}^u/\,\|\mathbf{g}^u\|^2$ by Eq. (3). Therefore, it is required that the fourth derivative of the interpolating polynomial is used, being a quintic polynomial the lowest odd degree polynomials that can be used to formulate accurately the prescribed motion constraint.

5 Prescribed Motion Constraint Application Cases

In order to study the use of different piecewise interpolation methodologies, on a perspective of prescribed motion constraint violation, two curves, associated to the track centerlines designs of roller coasters, are considered: an horizontal ellipse with a corkscrew, and; a more complex three-dimensional track with a geometry analogous to a complete roller coaster. These curves are parameterized using three different interpolation approaches: cubic spline 'Spline 3'; shape preserving cubic hermit polynomial 'Pchip' The MathWorks Inc (2002); quintic spline 'Spline5'.

The use of the prescribed motion constraint is demonstrated by enforcing that a vehicle represented by a single rigid body follows the defined curves. The orientation of the Frenet frame associated to the curves is rotated about the tangent vector in each particular geometry by an angle φ, as shown in Fig. 5, to obtain the desired cant angle that corresponds to the equilibrium cant for the vehicle travelling at a prescribed speed. Details on the application of the cant definition with reference

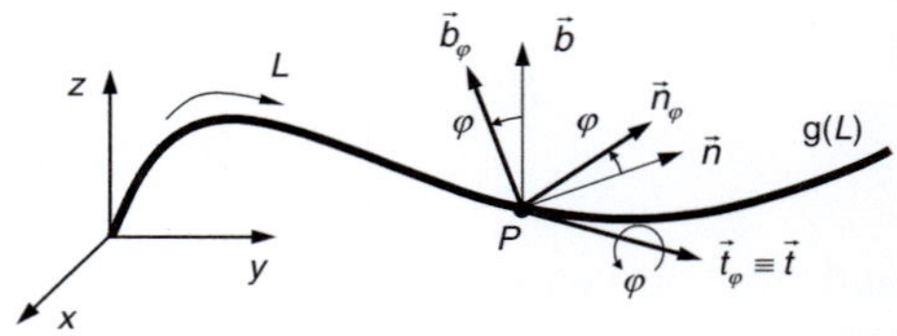

Fig. 5 Contribution of cant angle to the track model

to the equilibrium cant can be found in the work by Pombo and Ambrosio (2003), Pombo and Ambrósio (2007).

The dynamic analyzes of the two cases presented here are performed with all polynomial interpolators listed and with the application of the Baumgarte stabilization for the correction of the kinematic constraint violations. Note that only a constraint stabilization method is used and that constraint correction procedures, such as the coordinate partition method, are not considered in this work.

5.1 Ellipse with Corkscrew

The ellipse track, presented on Fig. 6, starts with a straight entryway designed to allow a smooth entry on the ellipse part of the path having a gradual change of the cant angle in order to match that of the ellipse at junction. The ellipsoidal segment of the track has a cant angle that corresponds to the equilibrium cant for the speed at which the vehicle travels. Taking the central part of the ellipse a two roll corkscrew segment is implemented by means of cant variation. The rigid body here considered starts with an initial velocity of 10 m/s and has a mass of 350 kg and inertias of $I_{\zeta\zeta}$= 50 Kg m^2 and $I_{\eta\eta}$=$I_{\xi\xi}$=120 Kg m^2.

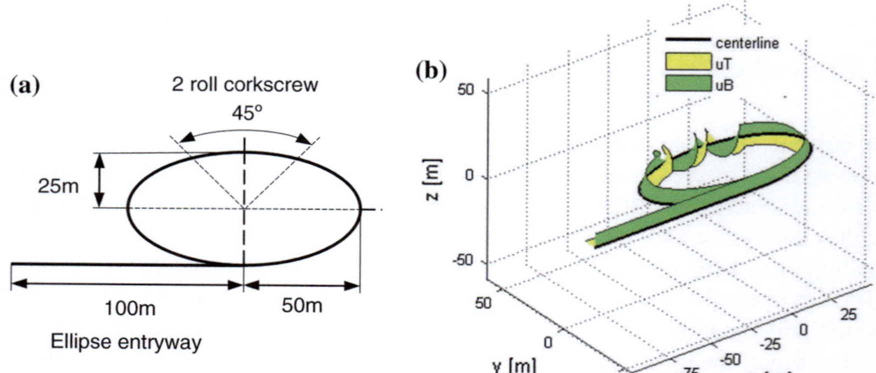

Fig. 6 Ellipse track characteristics: **a** geometric characteristics, **b** three-dimensional representation of the track *centreline* with a sweep of the unitary normal and binormal vectors

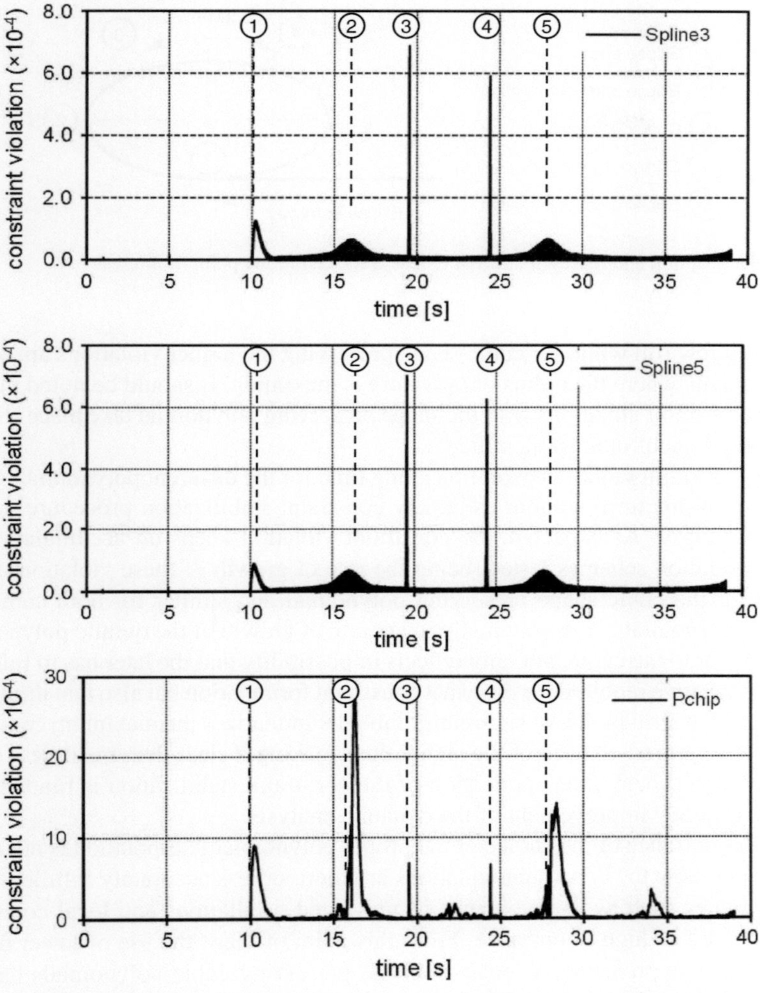

Fig. 7 Ellipse centreline constraint violation evolution along time with Baumgarte stabilisation for each interpolation polynomial

The evolution in time of the constraint violation value for each interpolation procedure considered in this work, when using the Baumgarte stabilization, is presented on Fig. 7. The constraint violation peaks, with all interpolating schemes, takeplace when the vehicle passes in some particular points of the track identified by ① through ⑤ and described on Fig. 8.

The results show that the cubic and quantic polynomials lead to similar evolutions of the constraint violations while the cubic shape preserving polynomial presents constraint violations one order of magnitude higher. The maximum constraint violations observed with the cubic and quantic polynomials take place at the start and end of

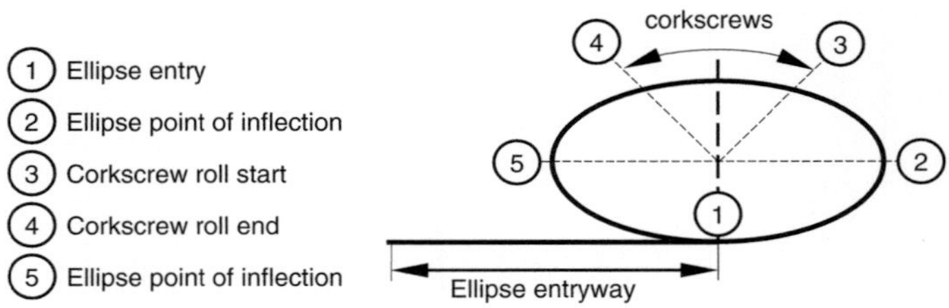

Fig. 8 Description and representation of ellipse track centreline point of interest

the corkscrew roll while for cubic shape preserving the higher violations are for the ellipse points where the radius of curvature is maximum. It should be noted that the higher constraint violations with the shape preserving polynomial take place slightly after the points of interest identified.

The constraint violation evolution along time for the different polynomials tested here, but in this turn, without using any constraint stabilization procedure, is presented in Fig. 9. As expected, the constraint violation keeps on accumulating for all interpolation schemes tested, being the rate of growth of these violations much higher for the cubic shape preserving polynomial and similar for both cubic and quintic polynomials. The marginally lower rate of growth of the quintic polynomial, with respect to the cubic, not only reflects impossibility that the later has to fulfil the fourth derivative required for the exact constraint formulation but also that the contribution of that term is almost negligible. Table 1 summarizes the maximum constraint violation observed on each of the simulations making it clear that, regardless of the interpolation scheme, the application of the constraint stabilization is fundamental for the accuracy improvement of the dynamic analysis.

The importance of the choice of the proper polynomial interpolation is not extinguished on how the constraint equations are more or less accurately fulfilled. Geometric issues such as the avoidance of unwanted oscillations and local control of the curve are of high importance. From this point of view, the use of lower degree polynomials is preferred. The use of shape preserving cubic polynomials handles both the geometric requirements for continuity and the local control avoiding overshooting. The IMSL Fortran routine `CSCON` (Visual NumericsInc 1997), based on the work by Irvine et al. (1986), provided good results with low constraint violations (Pombo and Ambrosio 2003; Pombo and Ambrósio 2007). However, the Matlab function `Pchip` (The MathWorks Inc 2002) used here, based on the work of Fritsch and Carlson (1980), does not provide an acceptable accuracy for this application.

Table 1 Ellipse maximum constraint violation for each interpolation polynomial.

	Maximum constraint violation		
	Spline 3	Spline 5	Pchip
Stabilized	6.88×10^{-04}	7.01×10^{-04}	29.1×10^{-04}
No Stabilization	4.83×10^{-02}	3.28×10^{-02}	73.3×10^{-02}

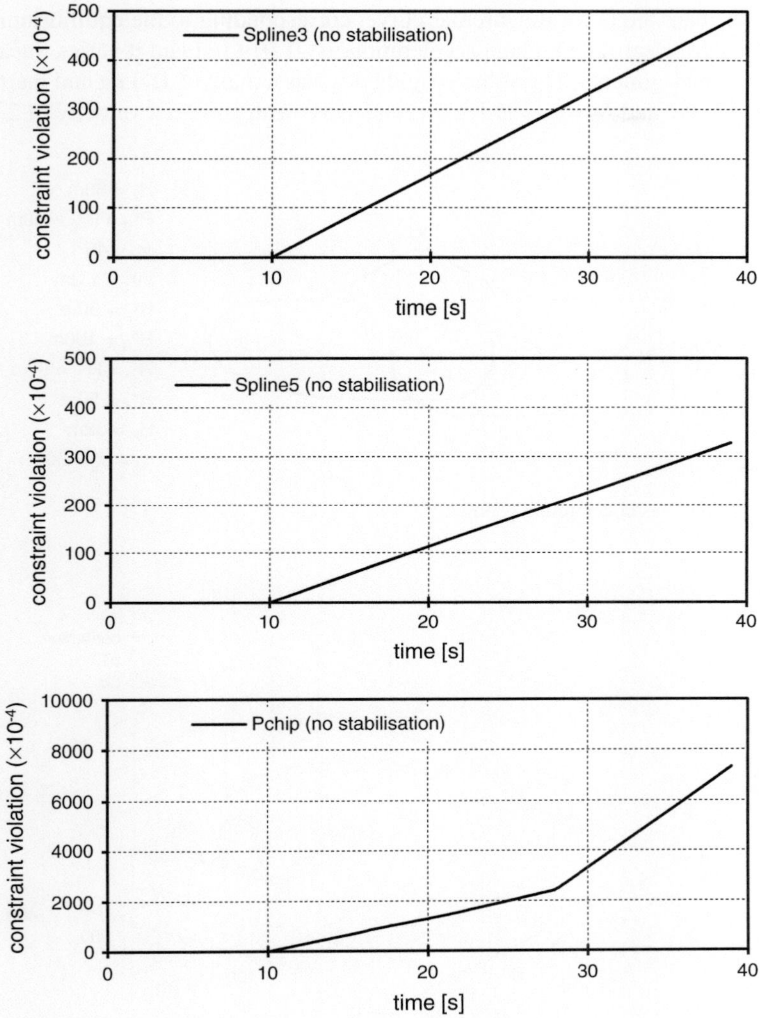

Fig. 9 Evolution of the *ellipse centreline* constraint violation for the different interpolating polynomials when no constraint stabilisation procedure is used

5.2 Roller Coaster

To demonstrate the performance of the polynomial interpolation schemes in a more complex and general geometry a roller coaster track, illustrated on Figs. 10 and 11, is used here. The roller coaster model, initially presented by Pombo and Ambrosio (2003), Pombo and Ambrósio (2007), has the geometry shown in Fig. 10 being the transition into and out of the curve G–I modeled with parabolic transition curves of

60 m. The cant angle for the circular curve, corresponding to the equilibrium cant, i.e. the cant for zero track plane acceleration, is –1.014 rad and it varies linearly in the transition segments. The vehicle rigid body has a mass of 150 kg and inertias of $I_{\zeta\zeta}$= 25 Kg.m^2 and $I_{\eta\eta}$=$I_{\xi\xi}$=65 Kg.m^2, and starts with an initial velocity of 2 m/s.

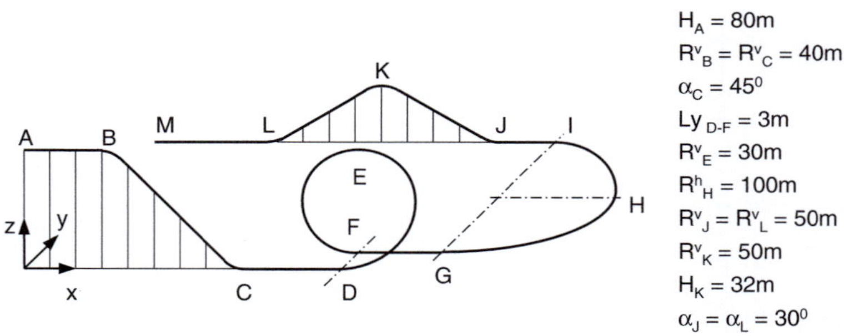

Fig. 10 Roller coaster centreline geometry

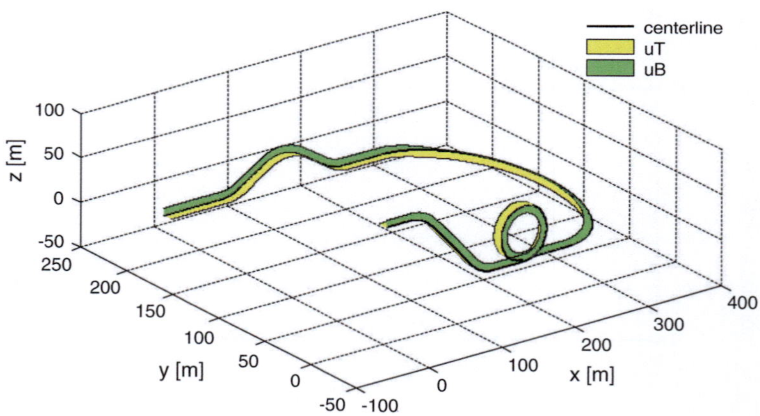

Fig. 11 Three-dimensional representation of the track centreline including a sweep of the unitary normal and binormal vectors

The evolution of the constraint violation with each one of the interpolation polynomials is presented in Fig. 12, for the cases in which the Baumgarte constraint stabilization is used, and in Fig. 13, for the cases simulated without constraint violation stabilization. As for the simpler case presented in the previous section, the constraint stabilization procedure plays a central role on the accuracy of the dynamic analysis, regardless of the interpolation scheme selected. Also, just as before, the cubic shape preserving polynomial `Pchip` performance is not satisfactory, regardless of the improvements observed when using the constraint stabilization.In this

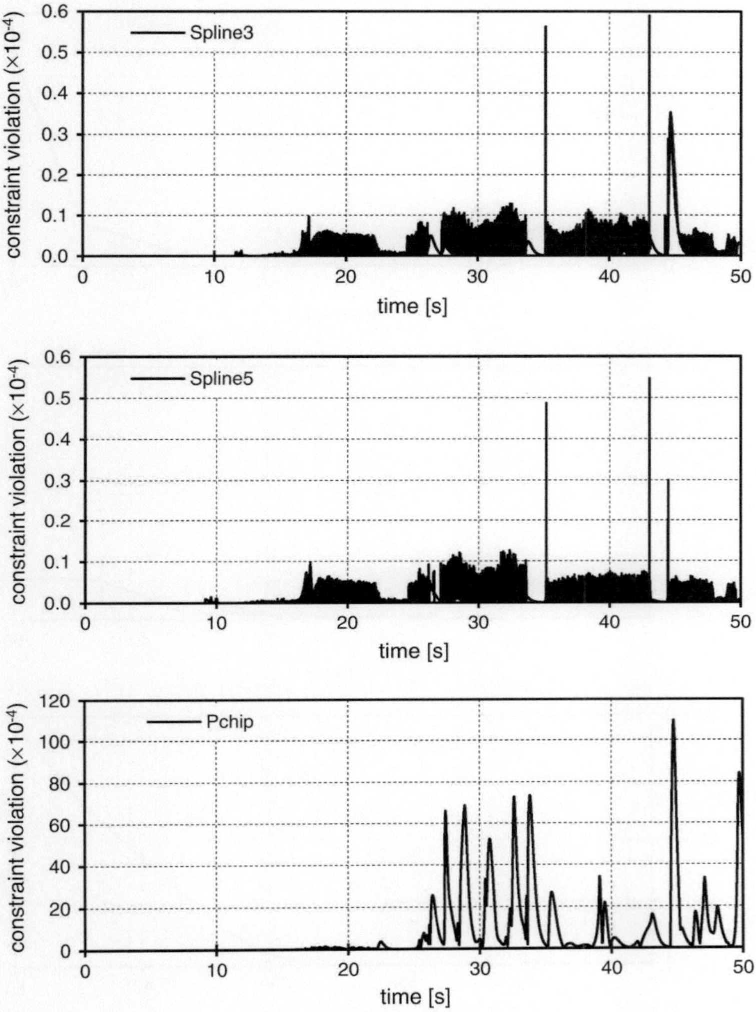

Fig. 12 Evolution of the roller coaster centreline constraint violation, for each piecewise interpolation methodology, with Baumgarte constraint stabilisation

general geometry it is observed that the quantic polynomial leads to an accumulation of the constraint violations at a lower rate than what is observed with the cubic polynomial, being overall one order of magnitude lower.

A summary of the maximum constraint violation evaluated on each of the simulations for the roller coaster track is presented on Table 2. Although, overall, these results have the same trends already observed for the ellipse track, the difference in performance between the quantic and cubic polynomials, for dynamic analysis

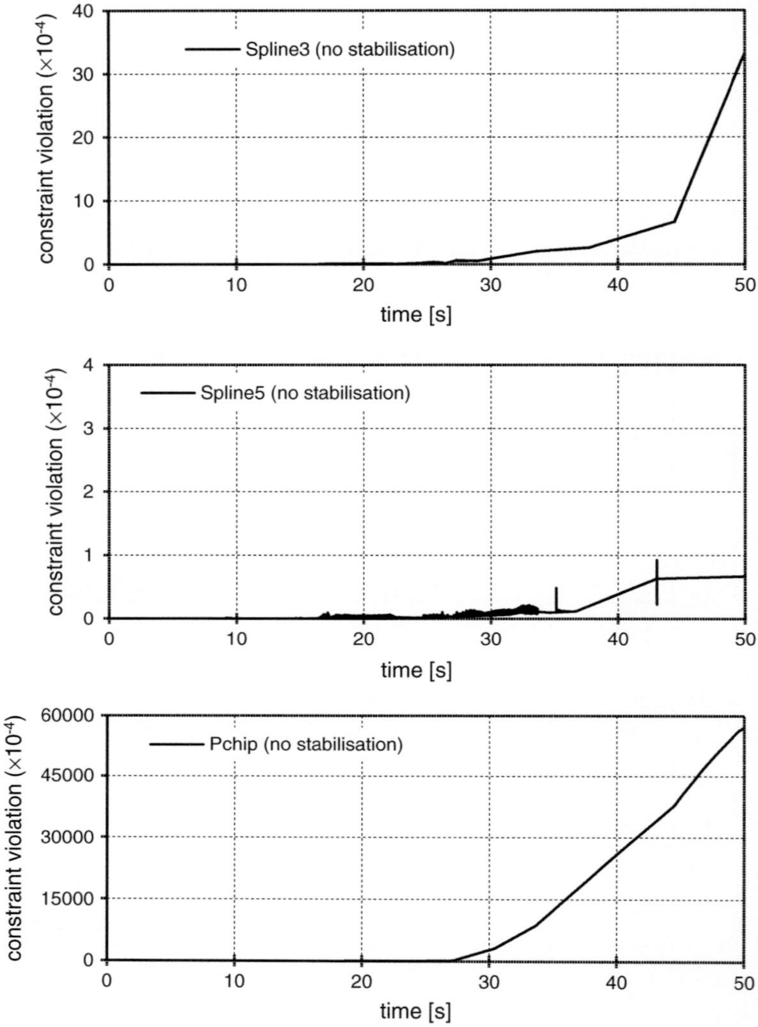

Fig. 13 Evolution of the roller coaster centreline constraint violation for each piecewise interpolation methodology when no constraint stabilisation is used

without constrain stabilization, is clearer now. This case suggests that the role played by the fourth derivative of the curve equation with respect to its arc length, as required in Eq. (18) plays a non-neglectable role. However, because the use of the constraint stabilization procedure in the framework of multibody dynamic formulations, with dependent coordinates, is unavoidable the differences of accuracy and performance between the cubic and quantic polynomials vanish.

Table 2 Roller coaster maximum constraint violation for each interpolation polynomial

	Maximum constraint violation		
	Spline 3	Spline 5	Pchip
Stabilized	5.90×10^{-05}	5.47×10^{-05}	1.10×10^{-02}
No Stabilization	3.33×10^{-03}	9.28×10^{-05}	$5.71\times10^{+00}$

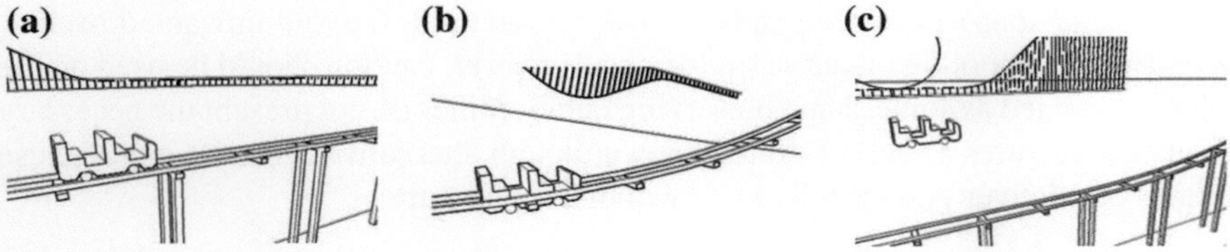

Fig. 14 Visual representation of a constraint violation: **a** No violation, **b** Penetration between track and vehicle, **c** Separation between track and vehicle

It must be noticed that, in general dynamic analysis of multibody systems, the constraint violations are reflected in the dynamic response in much more than the stability of the complete process. The physical interpretation of the constraint violations, for the roller coaster case is illustrated in Fig. 14 for several values of the case in which the cubic shape preserving spline is used. Usually the existence of the constraint violations have the same effect as spurious forces in the system being responsible not only for lack of accuracy of the system dynamic response but also for slowing the integration due to the need for a reduction of the time step of variable time step integrators. For these reasons, constraint stabilizations procedures and/or constraint correction methods must always be used.

6 Conclusions

The formulation of a prescribed motion constraint requires the parameterization of a curve with a polynomial with an order that fulfils the geometric requirements of smoothness, shape control and an accurate description of the kinematic constraint and of its second time derivative. From the geometric point of view C^2 polynomials would suffice but to obtain the accurate formulation of the constraint the fourth order is the minimum required. However, regardless of the order of the polynomial actually used for the prescribed motion constraint, when dependent coordinates, such as Cartesian coordinates, are used in the multibody dynamic formulation the small numerical errors always present in the numerical methods used to solve and integrate the equations of motion tend to accumulate, ultimately leading to the violation of the kinematic constraints of the system. Therefore, the use of constraint stabilization or correction methods is unavoidable.

This work shows that the constraint stabilization methods effectively stabilize the numerical errors and also those resulting from the use cubic interpolating polynomials in the constraint formulation making the results indistinguishable from those obtained with quintic polynomials. Due to the fact that both cubic and quintic polynomial have the required degree of smoothness to describe geometric problem and that the cubic polynomials have better local control properties it is shown that they are more advantageous in the formulation of the prescribed motion constraint. The use of cubic shape preserving splines may be required to avoid unwanted oscillations and overshooting on curve transitions. However, caution should be used on the scheme selected as some shape preserving cubic splines do not present the necessary geometric features as demonstrated here, although alternative formulations of these splines can deliver good results as shown in the literature.

Acknowledgments To Prof. Andrés Kecskeméthy for the challenges and discussions throughout the years and to Dr. Martin Tandl, whose work is a reference to roller-coaster design.The work reported here was possible due to the funding by FCT (Foundation for Science and Technology) under the projects SMARTRACK (PTDC/EME-PME/101419/2008) and WEARWHEEL (PTDC/EME-PME/115491/2009).

References

Baumgarte J (1972) Stabilization of constraints and integrals of motion in dynamical systems. Comput Methods Appl Mech Eng 1:1–16

Bayo E, Avello A (1994) Singularity-free augmented Lagrangian algorithms for constrained multibody dynamics. Nonlinear Dyn 5(2):209–231

Darboux G (1889) Leconssur la theorieGenerale des surfaces et les applications geometriques du calcul infinitesimal, vol 2. Gauthier-Villars, Paris, France

De Boor C (1978) A practical guide to splines. Springer, New York

Farin GE (1990) Curves and surfaces for computer aided geometric design: a practical guide, 2nd edn. Academic Press, Boston

Frenet JF (1852) Sur les courbes a double courbure. J de MathemathiquesPuresetAppliquees. 17:437–447

Fritsch FN, Carlson RE (1980) Monotone piecewise cubic interpolation. SIAM J Numer Anal 17(2):238–246

Gear CW (1971) Simultaneous numerical solution of differential-algebraic equations. IEEE Trans Circ Theor 18(1):89–95

Irvine LD, Marin SP, Smith PW (1986) Constrained interpolation and smoothing. Constr Approx 2:129–151

de Jalon J, Bayo E (1993) Kinematic and dynamic simulation of multibody systems. Springer, Heidelberg

Neto MA, Ambrósio J (2003) Stabilization methods for the integration of DAE in the presence of redundant constraints. Multibody Syst Dyn 10:81–105

Nikravesh PE (1988) Computer-aided analysis of mechanical systems. Prentice-Hall, Englewood Cliffs, New Jersey

Pombo J, Ambrosio J (2003) General spatial curve joint for rail guided vehicles: kinematics and dynamics. Multibody Syst Dyn 9:237–264

Pombo J, Ambrósio J (2007) Modelling tracks for roller coaster dynamics. Int J Veh Des 45(4):470–500

Shampine L, Gordon M (1975) Computer solution of ordinary differential equations. Freeman, San Francisco
Tandl M (2008) Dynamic simulation and design of roller coaster motion, Ph.D. Dissertation, Universitat Duisburg-Essen, Duisburg, Germany
Tandl M, Kecskemethy A (2006) Singularity-free trajectory tracking with Frenet frames. In: Husty M, Schroeker H-P (eds) Proceedings of the 1st conference EuCoMeS. Obergurgl, Austria
The MathWorks Inc (2002) Using MATLAB, TheMathWorks Inc., Natick, Massachussetts
Visual NumericsInc (1997) IMSL Fortran 90 Math Library 4.0—Fortran subroutines for mathematical applications. Huston, Texas

Author Index

A. Kecskeméthy and F. Geu Flores (eds.), *Interdisciplinary Applications of Kinematics*,
Mechanisms and Machine Science 26, DOI 10.1007/978-3-319-10723-3

MIX
Papier aus verantwortungsvollen Quellen
Paper from responsible sources
FSC® C105338

If you have any concerns about our products,
you can contact us on
ProductSafety@springernature.com

In case Publisher is established outside the EU,
the EU authorized representative is:
Springer Nature Customer Service Center GmbH
Europaplatz 3, 69115 Heidelberg, Germany

Printed by Libri Plureos GmbH
in Hamburg, Germany